Ethics in Health

Sam Sparrow

ISBN: 978-1-77961-879-5
Imprint: Telephasic Workshop
Copyright © 2024 Sam Sparrow.

Contents

Introduction to Health and Wellness Ethics

The Importance of Ethics in Health and Wellness

Defining Ethics in the Context of Health and Wellness

In order to navigate the complex and challenging landscape of health and wellness, it is essential to have a solid understanding of the ethical principles that guide decision-making in this field. Ethics provides a framework for evaluating what is right and wrong, just and unjust, and helps us to grapple with moral dilemmas that arise in healthcare and wellness settings. In this section, we will explore the definition of ethics in the context of health and wellness, and how it relates to the promotion of overall well-being and the delivery of healthcare services.

What is Ethics?

Ethics is a branch of philosophy that deals with moral principles and values that shape human behavior and decision-making. It aims to answer questions about what is morally right or wrong, good or bad, and just or unjust. In the context of health and wellness, ethics provides a framework to evaluate how health-related decisions and actions impact individuals, communities, and societies as a whole.

Ethics is concerned with examining human values and applying critical reasoning to address complex moral issues. It helps to ensure that healthcare and wellness professionals uphold a set of moral standards when delivering care, making decisions, and promoting health and well-being.

1

Ethics in Health and Wellness

Ethics plays a central role in the field of health and wellness, as it encompasses various aspects of healthcare delivery, public health, and personal well-being. It guides professionals in their interactions with patients, as well as their obligations towards society. The ethical principles in health and wellness serve to protect the rights and welfare of individuals, promote justice and equity, and ensure that decisions made in healthcare are both ethical and morally sound.

Ethical considerations in health and wellness involve weighing the potential benefits and harms of interventions, respecting autonomy and informed consent, upholding principles of non-maleficence and beneficence, ensuring justice and equity in resource allocation, maintaining privacy and confidentiality, and promoting dignity and respect for all individuals.

Key Ethical Principles in Health and Wellness

Several key ethical principles provide a foundation for ethical decision-making in the context of health and wellness. These principles serve as guiding values that help healthcare professionals navigate moral dilemmas and promote the well-being of individuals and communities. Let's explore some of these ethical principles:

1. **Autonomy and Informed Consent:** Autonomy refers to an individual's right to make decisions regarding their own health and well-being. Informed consent is the process through which individuals are provided with relevant information about their healthcare options in order to make autonomous decisions. Respecting autonomy and obtaining informed consent are crucial ethical considerations in healthcare and wellness, as they uphold individuals' rights and values.

2. **Beneficence and Non-Maleficence:** Beneficence refers to the duty to promote the well-being of others and to act in their best interests. Non-maleficence, on the other hand, requires healthcare professionals to do no harm and to minimize potential risks associated with interventions. Balancing these principles ensures that healthcare and wellness interventions provide maximum benefit with minimal harm.

3. **Justice and Equity:** The principles of justice and equity focus on fairness in the distribution of healthcare resources and access to services. It requires that healthcare and wellness systems strive for equal treatment and opportunities for all individuals, regardless of their socio-economic status, gender, race, or other social determinants of health.

4. **Privacy and Confidentiality**: Privacy refers to an individual's right to keep personal information confidential and protected from unauthorized access. Confidentiality, on the other hand, refers to the duty of healthcare professionals to maintain the privacy of patient information. Respecting privacy and confidentiality is essential in building trust between healthcare providers and patients, and in maintaining the autonomy and dignity of individuals.

5. **Dignity and Respect**: Dignity and respect involve recognizing every individual's inherent worth and treating them with compassion and fairness. It requires healthcare professionals to acknowledge and honor the diversity of values, beliefs, and cultural backgrounds of patients, and to avoid discrimination or stigmatization.

It is important to note that these ethical principles are interconnected and often require a delicate balance in real-world situations. They provide a framework for healthcare professionals to navigate complex moral dilemmas and ensure that decisions and actions promote the best possible outcomes for individuals and communities.

Applying Ethics in Health and Wellness

Applying ethics in health and wellness involves recognizing and critically evaluating ethical issues that arise in various contexts, including clinical practice, public health interventions, research, policy development, and resource allocation. Professionals should strive to uphold ethical principles and engage in reflective practice to continually improve ethical decision-making.

To apply ethics effectively in health and wellness, professionals need to develop ethical reasoning skills, cultural competence, and an understanding of social determinants of health. They should actively engage in ethical discussions, seek guidance from ethical codes and guidelines, and collaborate with interdisciplinary teams to ensure that ethical considerations are incorporated into decision-making processes.

Conclusion

Defining ethics in the context of health and wellness is essential for understanding the moral principles and values that guide decision-making in this field. It encompasses a range of ethical issues, from individual autonomy and informed consent to justice and equity in resource allocation. By embracing and applying the core ethical principles, healthcare professionals can navigate complex moral dilemmas, promote the well-being of individuals and communities, and contribute to a more equitable and just health and wellness future.

Key Takeaways:

+ Ethics provides a framework for evaluating what is right and wrong, just and unjust in the context of health and wellness.

+ Ethical principles in health and wellness include autonomy, beneficence, non-maleficence, justice, privacy, confidentiality, dignity, and respect.

+ Applying ethics in health and wellness involves recognizing ethical issues, engaging in reflective practice, and incorporating ethical considerations into decision-making processes.

Resources:

+ American Medical Association. (2016). Code of Medical Ethics.

+ World Health Organization. (2002). Ethical considerations in public health interventions: A guidance document.

+ Gillon, R. (2003). Medical ethics: Four principles plus attention to scope. BMJ, 322(7294), 848.

Historical Overview of Ethics in Health and Wellness

Ethics in health and wellness have been a topic of discussion for centuries, as societies grapple with the moral dilemmas and responsibilities surrounding the provision of healthcare services and the promotion of overall well-being. This historical overview will provide some key insights into the development of ethical considerations in this field, tracing back to early philosophical and medical traditions.

Ancient Philosophical Foundations

The foundations of ethics in health and wellness can be traced back to ancient philosophical traditions in Greece and other civilizations. The Greek philosopher Hippocrates, often referred to as the father of medicine, made significant contributions to the ethical principles of healthcare. He emphasized the importance of the physician's moral character and the ethical duty to prioritize the patient's well-being. The Hippocratic Oath, a code of conduct for physicians, outlined the principles of medical ethics, including the duty to do no harm (non-maleficence) and to maintain patient confidentiality.

Similarly, the teachings of ancient philosophers such as Socrates, Plato, and Aristotle explored ethical concepts that were relevant to healthcare. Plato's dialogues, for instance, discussed the ideas of justice, virtue, and the pursuit of the highest good. Aristotle's concept of virtue ethics, which focuses on developing virtuous character traits, also provided a framework for understanding moral decision-making in healthcare.

Religious and Spiritual Influences

Religious and spiritual beliefs have played a significant role in shaping ethical considerations in health and wellness throughout history. In many societies, religious texts and teachings provided guidance on principles such as compassion, justice, and the sanctity of life. For example, the Christian tradition emphasized the importance of caring for the sick, promoting the idea of healing as a divine calling.

In addition to Christianity, other religious and spiritual traditions, such as Islam, Judaism, Hinduism, and Buddhism, have also contributed to ethical deliberations in health and wellness. These traditions often emphasize the interconnectedness of mind, body, and spirit, highlighting the importance of holistic approaches to well-being.

Bioethical Milestones

The field of bioethics emerged in the mid-20th century as advancements in medical technology raised new ethical challenges. The atrocities committed during World War II, such as the Holocaust and medical experiments conducted by Nazi doctors, led to a heightened awareness of the need for ethical guidelines in medical research and treatment.

In 1947, the Nuremberg Code was established in response to the unethical experiments conducted by Nazi doctors. It laid out principles for ethical research involving human subjects, including the requirement for voluntary informed consent and the importance of minimizing risks to participants.

In the 1960s and 1970s, further milestones in bioethics were achieved. The case of Henrietta Lacks, an African American woman whose cells were used for scientific research without her knowledge or consent, highlighted the need for patient autonomy and informed consent. This case also raised important questions about racial and social justice in healthcare.

The development of reproductive technologies, such as in vitro fertilization (IVF), raised ethical concerns about the beginning of life and the right to procreate. These issues spurred debates on the moral status of embryos and the rights of potential parents.

Contemporary Debates

In recent years, ethical considerations in health and wellness have expanded to include a wide range of topics. Some of the key contemporary debates include the ethics of genetic testing and screening, end-of-life care and euthanasia, global health inequalities, and the impact of technology on healthcare.

The rapid advancement of genetic technologies has raised ethical questions regarding privacy, discrimination, and the potential for eugenics. Issues such as prenatal genetic testing and genetic engineering have sparked discussions on the limits of human intervention in the natural course of life.

The debates around end-of-life care have become increasingly prominent, with discussions on physician-assisted death, the right to die with dignity, and the provision of palliative care. These topics touch on deeply personal beliefs, cultural norms, and the balancing of individual autonomy with societal well-being.

Global health inequalities continue to be a significant ethical concern, with debates on access to healthcare, the distribution of resources, and the responsibilities of affluent nations in addressing global health crises. These discussions highlight the importance of justice, equity, and human rights in healthcare systems.

Lastly, the rise of technology in healthcare, such as telemedicine, artificial intelligence, and personal health monitoring devices, brings forth ethical dilemmas related to privacy, data security, and the potential for bias in algorithmic decision-making.

In conclusion, the historical overview of ethics in health and wellness demonstrates the evolution of ethical considerations in healthcare from ancient philosophical foundations to contemporary bioethical debates. Understanding this historical context is vital for navigating the moral dilemmas, promoting justice, and shaping a more equitable, just, and sustainable world in the field of health and wellness.

Resources:

- Beauchamp, T. L., & Childress, J. F. (2019). Principles of biomedical ethics. Oxford University Press. - Jonsen, A. R., Siegler, M., & Winslade, W. J. (2015). Clinical ethics: A practical approach to ethical decisions in clinical medicine (8th ed.). McGraw-Hill Education. - Macklin, R. (2003). Bioethics, vulnerability, and protection. Bioethics, 17(5-6), 472-486. - Pellegrino, E. D., & Thomasma, D. C. (2005). The virtues in medical practice. Oxford University Press.

Ethical Theories and Approaches to Health and Wellness

In order to navigate the complex moral dilemmas surrounding health and wellness, it is important to have a solid understanding of the ethical theories and approaches that can guide our decision-making process. This section will provide an overview of some key ethical frameworks that can be applied to the field of health and wellness.

Utilitarianism

Utilitarianism is an ethical theory that emphasizes the maximization of overall happiness and well-being. According to this framework, the moral action is the one that produces the greatest amount of happiness for the greatest number of people. In the context of health and wellness, a utilitarian approach would involve making decisions that prioritize the well-being of the majority over the interests of individuals or smaller groups.

For example, when making decisions about resource allocation in healthcare, a utilitarian perspective would suggest directing resources towards interventions that have the greatest potential to improve the overall health and well-being of the population. This might involve prioritizing preventive measures or treatments for conditions that have a significant impact on public health.

However, a critique of utilitarianism in the context of health and wellness is that it may neglect the needs and interests of marginalized or vulnerable populations. For instance, certain interventions might benefit the majority but not address the specific needs and concerns of minority groups or individuals with unique healthcare requirements.

Deontology

Deontology is an ethical theory that focuses on the inherent duty or moral obligation of individuals. According to deontological principles, some actions are inherently right or wrong, regardless of their consequences. This theory emphasizes principles such as autonomy, dignity, fairness, and respect for individuals.

In the realm of health and wellness, deontological approaches highlight the importance of respecting individual autonomy and promoting informed consent. It emphasizes the right of individuals to make decisions about their own health, with the understanding that healthcare providers should act in their patients' best interests and respect their preferences.

For example, deontological ethics would suggest that healthcare providers have an obligation to obtain informed consent from patients before performing medical procedures or starting treatments. This ensures that individuals have the necessary information to make autonomous decisions about their own healthcare.

However, a challenge with deontology lies in defining what is inherently right or wrong in a given situation. Different ethical principles may come into conflict, requiring careful consideration and balancing of competing duties.

Virtue Ethics

Virtue ethics focuses on the character traits and virtues that contribute to the development of a good and virtuous person. Rather than solely focusing on rules or consequences, virtue ethics emphasizes the cultivation of virtuous habits and moral virtues.

In the context of health and wellness, virtue ethics directs attention towards the character traits and virtues that are important for healthcare providers. These virtues might include compassion, empathy, integrity, and honesty. By embodying these virtues, healthcare professionals can contribute to the well-being of their patients and promote ethical healthcare practices.

For instance, a healthcare provider who embodies the virtue of compassion would prioritize patient-centered care and strive to alleviate suffering. This could involve taking the time to listen to patients' concerns, empathizing with their experiences, and providing support in challenging medical situations.

Virtue ethics complements other ethical frameworks by emphasizing the importance of character and personal development. It encourages healthcare professionals to strive towards becoming morally excellent individuals who can make ethically sound decisions in various healthcare contexts.

Principlism

Principlism, also known as ethical pluralism, is an approach to ethics that emphasizes the use of foundational ethical principles to guide decision-making. The four commonly accepted principles in healthcare ethics are autonomy, beneficence, non-maleficence, and justice.

Autonomy refers to respecting individuals' right to make decisions about their own healthcare. Beneficence involves promoting the well-being and welfare of others. Non-maleficence requires avoiding harm or minimizing the negative impact of actions. Justice entails ensuring fairness and equality in the distribution of healthcare resources and access to healthcare.

Principle-based ethics provides a comprehensive framework to evaluate and address various ethical dilemmas that arise in health and wellness. By considering these principles, healthcare professionals can navigate challenging situations and strive for ethical decision-making.

For example, when faced with a difficult ethical decision, a healthcare provider might consider the principle of autonomy by respecting a patient's preferences and choices. At the same time, they may also consider the principle of beneficence by assessing the potential benefits and harms of different treatment options.

It is important to note that these principles may come into conflict with each other, requiring thoughtful consideration and ethical reasoning to strike a balance.

Feminist Ethics

Feminist ethics provides a unique perspective on ethical issues in health and wellness by emphasizing the importance of gender and intersectionality. This approach recognizes that gender, along with other social identities, influences experiences of health, access to healthcare, and power dynamics within healthcare systems.

Feminist ethics calls for a critical examination of social and institutional structures that perpetuate gender inequalities in health. It also highlights the importance of incorporating diverse voices and perspectives in the decision-making process.

For example, feminist ethics would critique healthcare policies and practices that marginalize or neglect the specific health needs of women, transgender individuals, and other gender minorities. It advocates for equitable and compassionate care that takes into account the unique experiences and concerns of different gender groups.

Feminist ethics also aims to challenge power imbalances within healthcare systems and promote inclusivity and social justice. By recognizing and addressing systemic gender biases, healthcare professionals can work towards creating more just and equitable healthcare environments.

Conclusion

Ethical theories and approaches provide a valuable foundation for navigating the complex moral dilemmas that arise in the field of health and wellness. Utilitarianism, deontology, virtue ethics, principlism, and feminist ethics offer different perspectives and frameworks for ethical decision-making.

No single ethical theory or approach is perfect or universally applicable, and each framework has its own strengths and limitations. By considering and integrating various ethical perspectives, healthcare professionals can adopt a more holistic and comprehensive approach to addressing ethical challenges in health and wellness. It is important to exercise critical thinking and ethical reasoning, taking into account the values, principles, and needs of individuals and communities when making informed ethical decisions.

Ethical Principles in the Delivery of Health and Wellness Services

Autonomy and Informed Consent

In the context of health and wellness, autonomy refers to an individual's right to make decisions about their own body, health, and well-being. It is the foundation of modern bioethics and plays a crucial role in the delivery of health and wellness services. Autonomy recognizes and respects the individual's ability to make choices based on their values, beliefs, and goals. Informed consent, on the other hand, is the process through which individuals are provided with relevant information about a proposed treatment or intervention, understand the potential risks and benefits, and voluntarily agree to or refuse the proposed course of action.

The principle of autonomy dates back to the ancient Greek philosophy, where the concept of self-governance was highly valued. However, the recognition of autonomy as a fundamental principle in healthcare ethics didn't emerge until the 20th century. Informed consent, as we understand it today, originated in response to unethical medical experiments carried out during World War II, such as the Tuskegee Syphilis Study and Nazi human experimentation. These cases highlighted the need to protect individuals' rights to make informed choices about their own bodies and health.

In healthcare practice, autonomy is closely linked to respect for persons. This principle requires healthcare providers to recognize and acknowledge the inherent worth, dignity, and rights of each individual. By promoting autonomy, healthcare professionals empower patients to actively participate in their healthcare decisions, fostering a patient-centered approach that values their perspectives, preferences, and values.

To ensure individuals can make autonomous decisions, healthcare providers must provide them with adequate information. This process is known as informed consent. Informed consent involves a meaningful dialogue between the healthcare provider and the patient, where the provider discloses information about the diagnosis, treatment options, potential risks and benefits, and alternatives. The patient, in turn, evaluates this information in light of their values and beliefs, making an informed decision about their healthcare.

It is important to note that informed consent is not a mere formality but a comprehensive process that respects the autonomy and dignity of individuals. The process must fulfill several key elements to be considered valid. These elements include:

+ **Capacity:** The patient must have the mental capacity to understand the information provided and make a decision. Healthcare providers have a responsibility to assess the patient's capacity and ensure they have the ability to comprehend the relevant information.

+ **Voluntariness:** The decision to provide informed consent should be free from coercion, manipulation, or undue influence. Patients should have the freedom to accept or reject a proposed treatment without fear of adverse consequences.

+ **Comprehension:** The information provided to the patient must be presented in a manner that they can understand. This requires healthcare providers to use clear and non-technical language, provide explanations, and address any questions or concerns the patient may have.

+ **Disclosure:** Healthcare providers have a duty to disclose all relevant information to the patient. This includes information about the nature of the condition, the proposed treatment or intervention, potential risks and benefits, alternative options, and the likely outcomes.

Ensuring informed consent can be a complex process, particularly in situations where the patient's capacity may be compromised, such as in emergency situations or when dealing with vulnerable populations. In such cases, healthcare providers may need to involve substitute decision-makers, such as family members or legal guardians, to make decisions on behalf of the patient while still considering their best interests.

Informed consent is not without its challenges. There may be instances where patients have limited health literacy or cultural differences that impact their understanding of medical information. In such cases, healthcare providers must strive to use culturally appropriate and plain language explanations, provide visual aids or educational materials, and engage interpreters when necessary to ensure that the patient fully comprehends the information.

Additionally, informed consent is not a one-time event but an ongoing process. As new information becomes available or treatment options change, healthcare providers must continue to communicate and engage with patients to ensure their understanding and continued informed decision-making.

In summary, autonomy and informed consent are essential components of ethical healthcare practices. Respecting patients' autonomy allows them to actively participate in decisions regarding their health and well-being, while informed consent ensures that they have the necessary information to make knowledgeable choices.   By upholding these principles, healthcare providers can promote patient-centered care and foster a relationship of trust and collaboration with their patients.

Key Principles of Autonomy and Informed Consent:

- Autonomy is the right of individuals to make decisions about their own health and well-being.

- Informed consent is the process through which individuals are provided with relevant information and voluntarily agree to or refuse a proposed treatment or intervention.

- Autonomy and informed consent promote patient-centered care and respect for persons.

- The elements of informed consent include capacity, voluntariness, comprehension, and disclosure.

- Informed consent is an ongoing process that requires clear communication, plain language explanations, and cultural sensitivity.

Case Study: Balancing Autonomy and Treatment Decisions

Consider the case of Sarah, a 40-year-old woman diagnosed with breast cancer. After extensive discussions with her healthcare provider, Sarah is presented with two treatment options: a lumpectomy followed by radiation therapy or a full mastectomy. Both options have similar rates of success in terms of cancer control, but the mastectomy carries a lower risk of recurrence.

Sarah, who highly values her body image and identifies strongly with her breasts, expresses a strong preference for the lumpectomy. She understands the potential risks and benefits of both options and acknowledges the possibility of recurrence. She believes that preserving her breasts will contribute positively to her overall well-being and self-confidence.

While the healthcare provider recommends the full mastectomy due to the lower risk of recurrence, they respect Sarah's decision. They acknowledge the importance of autonomy in the decision-making process and understand that

Sarah's psychological well-being is a significant factor. The provider ensures that Sarah has access to additional resources and support services to cope with her decision and any potential challenges that may arise.

This case demonstrates the complexity of balancing autonomy and treatment decisions. It highlights the importance of respecting patients' values, beliefs, and preferences, even when their choices may differ from healthcare providers' recommendations. By engaging in open and honest discussions and considering the holistic well-being of the patient, healthcare providers can navigate these ethical dilemmas and promote patient autonomy.

Beneficence and Non-Maleficence

In the realm of health and wellness ethics, two fundamental principles guide the actions of healthcare professionals: beneficence and non-maleficence. These principles form the basis of ethical decision-making and shape the way healthcare providers deliver care to their patients.

Understanding Beneficence

Beneficence refers to the ethical obligation of healthcare professionals to act in the best interest of their patients and promote their well-being. It requires healthcare providers to make choices that maximize the benefits to patients and minimize harm. The principle of beneficence is deeply rooted in the traditional principles of medicine, emphasizing the importance of physicians acting for the benefit of their patients.

To apply the principle of beneficence effectively, healthcare professionals must consider the potential risks and benefits of any intervention or treatment. They must also take into account the patient's values, preferences, and autonomy. In other words, beneficence should be balanced with respect for patient autonomy to ensure patient-centered and ethically sound decision-making.

Non-Maleficence: Do No Harm

Non-maleficence is closely related to beneficence and requires healthcare professionals to refrain from causing harm to their patients. The principle of non-maleficence is often summarized as "do no harm" and serves as a guiding principle in medical practice.

In the context of healthcare, avoiding harm means not only abstaining from causing physical harm but also preventing psychological, emotional, and social harm. Healthcare professionals should strive to mitigate any potential risks associated with medical interventions or treatments and prioritize patient safety.

Non-maleficence also extends beyond individual patients and encompasses the broader societal impact of healthcare decisions. Healthcare providers must consider the potential harm that could result from the dissemination of false or misleading information, the allocation of scarce resources, or the inequitable distribution of healthcare services.

Balancing Beneficence and Non-Maleficence

Balancing the principles of beneficence and non-maleficence can sometimes be a complex and challenging task. Healthcare professionals often face situations where interventions or treatments may have potential benefits but also carry significant risks. In such cases, a careful evaluation of the available evidence, informed consent, and shared decision-making with patients become crucial.

For example, consider the use of opioids for managing chronic pain. While opioids can provide significant pain relief for a patient, they also carry a risk of misuse, addiction, and overdose. Healthcare professionals must carefully assess the potential benefits and risks of opioid therapy, consider alternative treatments, and closely monitor patients for signs of opioid misuse.

To navigate this delicate balance, healthcare professionals should be equipped with up-to-date knowledge, critical thinking skills, and ethical decision-making frameworks. Ethical guidelines and professional codes of conduct can also provide valuable guidance in resolving dilemmas related to beneficence and non-maleficence.

Case Study: The Use of Antipsychotic Medications in Children

Let's consider a case study that highlights the ethical challenges associated with the principles of beneficence and non-maleficence. The use of antipsychotic medications in children with behavioral disorders can present such challenges.

On one hand, antipsychotic medications can effectively manage symptoms and improve the quality of life for children with severe behavioral disorders such as autism or attention-deficit/hyperactivity disorder (ADHD). The medications may help reduce aggression, impulsivity, and self-injurious behaviors, thus benefiting the child and their family.

However, antipsychotic medications also carry potential risks, including significant side effects such as weight gain, metabolic disturbances, and movement disorders. Furthermore, the long-term effects of these medications on children's brain development remain unclear.

In this case, healthcare professionals must carefully weigh the potential benefits of symptom relief against the risks associated with these medications. A comprehensive evaluation of the child's condition, consideration of alternative therapies, and full disclosure of potential risks to the child and their parents are essential.

Shared decision-making between healthcare providers, parents, and, if appropriate, the child is critical in reaching an ethically sound and patient-centered approach. The principle of beneficence requires healthcare professionals to prioritize the child's well-being, while the principle of non-maleficence mandates mitigating potential harms. Such a balanced approach ensures that the best interests of the child are considered while minimizing the risks involved.

Additional Resources

To further explore the principles of beneficence and non-maleficence in health and wellness ethics, consider the following resources:

- Veatch, R. M. (1987). "Doing no harm: The limits of risk." The Hastings Center Report, 17(2), 27-31.

- Beauchamp, T. L., & Childress, J. F. (2019). "Principles of Biomedical Ethics" (8th ed.). Oxford University Press.

- American Medical Association. (2016). Code of Medical Ethics: Opinions on Beneficence & Non-Maleficence. Retrieved from `https://www.ama-assn.org/delivering-care/ethics/` `beneficence-non-maleficence`

Remember, the principles of beneficence and non-maleficence should guide healthcare professionals in providing compassionate, effective, and ethically responsible care to their patients.

Justice and Equity

In the context of health and wellness, the principles of justice and equity play a crucial role in ensuring fair and equal access to healthcare services, resources, and opportunities. Justice refers to the fair distribution of benefits and burdens in society, while equity focuses on addressing disparities and promoting fairness.

Understanding Justice

Justice can be approached from different philosophical perspectives, such as utilitarianism, egalitarianism, and libertarianism. Utilitarianism emphasizes maximizing overall societal well-being, often through the allocation of resources based on need and effectiveness. Egalitarianism, on the other hand, promotes the equal distribution of resources and opportunities to ensure fairness. Libertarianism emphasizes individual autonomy, limiting the role of the state in redistributive efforts.

In the context of health and wellness, justice can be understood as ensuring that everyone has an equal opportunity to achieve and maintain good health. This implies addressing social determinants of health, such as income, education, housing, and access to healthcare, which can perpetuate health inequalities. Achieving justice in health requires recognizing and addressing systemic and structural factors that contribute to health disparities.

Health Disparities and Inequalities

Health disparities refer to differences in health outcomes and access to healthcare services among different population groups. These disparities are often linked to social determinants of health and can be influenced by factors such as race, ethnicity, socioeconomic status, gender, and geography. Addressing health disparities requires a focus on justice and equity.

Socioeconomic inequalities contribute significantly to health disparities. Individuals from lower-income backgrounds often face barriers to accessing quality healthcare services, preventive care, and health-promoting resources. They may also experience higher rates of chronic diseases and poorer health outcomes. Achieving health justice requires addressing income inequality and provide resources and opportunities for all individuals to attain optimal health.

Racial and ethnic disparities in health are another critical aspect of health justice. Historically marginalized communities face higher rates of chronic diseases, infant mortality, and limited access to healthcare services. These disparities can be attributed to factors such as systemic racism, discrimination, and limited cultural competence in healthcare delivery. Addressing health justice requires recognizing and rectifying these disparities by promoting culturally responsive care and addressing systemic racism.

Gender inequalities can also contribute to health disparities. Women, for example, may face barriers to accessing reproductive healthcare, gender-specific conditions, and may experience gender-based violence. LGBTQ+ individuals may also face unique health disparities due to discrimination and lack of understanding of their healthcare needs. Achieving health justice requires addressing these disparities and ensuring equal access to healthcare services for all genders and sexual orientations.

Promoting Health Justice and Equity

Promoting health justice and equity requires a multi-faceted approach that addresses the systemic and structural factors contributing to health disparities. Here are some strategies to promote health justice and equity:

1. Policy Interventions: Implementing policies that address social determinants of health, such as improving access to affordable housing, quality education, and nutritious food. Designing healthcare policies that prioritize the needs of marginalized and vulnerable populations can also contribute to health justice.

2. Health Equity Assessments: Conducting comprehensive assessments to identify and address health disparities within communities. This involves collecting data on social determinants of health, health outcomes, and access to healthcare services, in order to develop targeted interventions.

3. Culturally Competent Care: Ensuring healthcare providers are trained in cultural competency to provide appropriate care to diverse populations. This involves understanding and respecting the cultural beliefs, practices, and linguistic needs of patients.

4. Community Engagement: Engaging communities in decision-making processes related to health and wellness. This includes involving community members in the development and implementation of health programs to address their specific needs and challenges.

5. Education and Awareness: Promoting health education campaigns that address disparities and promote health justice. This involves raising awareness and providing accurate information about health conditions, preventive measures, and available resources.

6. Ethical Resource Allocation: Allocating healthcare resources in a fair and equitable manner. This involves considering the needs of different populations and prioritizing resource distribution based on health needs rather than socioeconomic status or other discriminatory factors.

7. Advocacy for Vulnerable Populations: Advocating for the rights and well-being of marginalized and vulnerable populations. This includes advocating for policies that protect their access to healthcare and address the social determinants of health that contribute to disparities.

It is important to recognize that promoting health justice and equity is an ongoing process that requires collaboration among policymakers, healthcare providers, communities, and individuals. By addressing social determinants of health and dismantling systemic barriers, we can work towards a more equitable and just health and wellness future for everyone.

Privacy and Confidentiality

Privacy and confidentiality are fundamental principles in the delivery of health and wellness services. They form the basis for establishing trust between healthcare providers and patients, as well as for protecting individuals' rights to control the disclosure of their personal health information. In this section, we will explore the importance of privacy and confidentiality in healthcare, the ethical considerations involved, and strategies for upholding these principles in practice.

Understanding Privacy in Healthcare

Privacy in healthcare refers to the right of individuals to control the access, use, and disclosure of their personal health information. It encompasses the protection of sensitive health data, as well as the individual's autonomy and ability to make informed decisions about their health. Privacy is crucial for maintaining trust between patients and healthcare providers, as it ensures that personal health information is kept confidential and used only for authorized purposes.

Confidentiality and its Role in Healthcare

Confidentiality goes hand in hand with privacy and relates to the duty of healthcare professionals to safeguard the personal health information shared by patients. It involves keeping information confidential and disclosing it only with the individual's consent or when required by law. Confidentiality is essential for maintaining the privacy rights of patients and promoting trust between healthcare providers and patients.

Ethical Considerations in Privacy and Confidentiality

Respecting privacy and maintaining confidentiality in healthcare practice raises several ethical considerations. Healthcare professionals must balance the individual's right to privacy and the need to share information in the best interest of the patient. Some key ethical considerations include:

+ **Informed consent:** Healthcare providers should obtain informed consent from patients before collecting, using, or disclosing their personal health information. This involves providing clear and understandable information about why the information is needed, how it will be used, and who will have access to it.

+ **Boundaries and confidentiality:** Healthcare professionals must establish and maintain appropriate boundaries to protect patient confidentiality. They should refrain from discussing patients' personal health information with unauthorized individuals and ensure that electronic health records and other health information systems are secure.

+ **Security and data breaches:** Healthcare providers are responsible for implementing robust security measures to protect patients' personal health information from unauthorized access, use, or disclosure. They should have protocols in place to promptly respond to and mitigate any data breaches that may occur.

+ **Legal and regulatory requirements:** Healthcare professionals must adhere to applicable laws and regulations regarding privacy and confidentiality. This includes compliance with health information privacy laws, such as the Health Insurance Portability and Accountability Act (HIPAA) in the United States, which sets standards for the protection of individuals' health information.

Challenges in Maintaining Privacy and Confidentiality

While privacy and confidentiality are essential in healthcare, several challenges exist in their maintenance. Technological advancements, such as electronic health records and telemedicine, have raised concerns about data security and unauthorized access. Additionally, the increasing reliance on data sharing for research and public health purposes has sparked debates about balancing individual privacy rights with societal benefits.

Strategies for Upholding Privacy and Confidentiality

To ensure the protection of privacy and confidentiality in healthcare, several strategies can be employed:

+ **Education and training:** Healthcare professionals should receive comprehensive education and training on privacy and confidentiality. This includes understanding legal and ethical obligations, best practices for data security, and protocols for handling confidential information.

+ **Technology safeguards:** Robust security measures, such as encryption and secure access controls, should be employed to protect patient information in electronic health records and other health information systems. Regular security audits and updates are necessary to address emerging threats.

+ **Policies and procedures:** Healthcare organizations should have clear policies and procedures in place to guide employees in the proper handling of confidential information. This includes protocols for obtaining informed consent, storing and sharing information securely, and responding to data breaches.

+ **Patient engagement:** Engaging patients in decisions regarding the use and disclosure of their personal health information can help promote trust and respect for their privacy rights. This can be achieved through informed consent processes, providing patients with access to their health records, and involving them in discussions about data sharing for research or public health purposes.

Case Study: Balancing Privacy and Public Health

Consider the case of a public health agency conducting a contact tracing program during a pandemic. The agency needs to collect and use individuals' personal health information to identify and notify close contacts of confirmed cases. While this program is vital for controlling the spread of the disease, it raises ethical considerations regarding privacy and confidentiality.

To balance privacy with the public health benefits of contact tracing, the agency can implement several measures. They can ensure that data collection is minimized to only collect essential information, such as names and contact details. The agency can also use de-identification techniques to remove personally identifiable information whenever possible.

Furthermore, the agency should establish strict access controls and protocols to ensure that only authorized personnel have access to the collected data. They should also inform individuals about the purpose of data collection, the safeguards in place to protect their information, and their rights regarding the use and disclosure of their data.

By taking these measures, the public health agency can safeguard the privacy and confidentiality of individuals while still carrying out vital public health activities.

Conclusion

Privacy and confidentiality play pivotal roles in the delivery of health and wellness services. Upholding these principles is essential for maintaining trust between healthcare providers and patients, respecting patients' autonomy, and protecting sensitive personal health information. By understanding the ethical considerations involved, implementing robust security measures, and engaging patients in decision-making, healthcare professionals and organizations can navigate the complexities of privacy and confidentiality in an increasingly interconnected and data-driven world.

Dignity and Respect

In the context of health and wellness ethics, the principles of dignity and respect are of paramount importance. Dignity refers to the inherent worth and value of every individual, regardless of their health status or circumstances. Respect, on the other hand, involves recognizing and honoring a person's autonomy, choices, and preferences. Together, these principles form the foundation for ethical decision-making and the delivery of compassionate and patient-centered care.

1. Dignity in Health and Wellness

Dignity is a fundamental aspect of human existence, and maintaining and promoting it is essential in healthcare settings. It involves acknowledging the uniqueness and individuality of each person, treating them with empathy and compassion, and upholding their rights and values.

One way to ensure dignity in healthcare is through the provision of person-centered care. This approach recognizes that patients are experts in their own lives and places them at the center of their care. It involves active listening, acknowledging their concerns, and involving them in decision-making processes.

Respecting patient privacy and confidentiality is another crucial aspect of maintaining dignity. Health professionals must handle sensitive information with utmost care and ensure that it is not disclosed without permission. This includes protecting electronic health records, maintaining confidentiality during consultations, and obtaining informed consent for the sharing of patient information.

2. Respect for Autonomy

Respect for autonomy is closely related to dignity and involves recognizing a person's right to make decisions about their own health and well-being. Autonomy requires that individuals have access to accurate information, understand the implications of their choices, and have the freedom to make decisions without coercion or outside influence.

Informed consent is a key ethical and legal requirement that promotes respect for autonomy. It is the process by which healthcare providers inform patients about the nature, risks, benefits, and alternatives of a particular treatment or intervention. Informed consent ensures that individuals can exercise their autonomy by giving their voluntary and informed agreement to a proposed course of action.

Respecting autonomy also means accommodating a person's values and preferences in the delivery of healthcare. This includes involving patients in shared decision-making, considering their cultural and religious beliefs, and tailoring interventions to meet their individual needs. Respecting autonomy may involve providing alternative treatment options or adapting care plans to align with the patient's goals and values.

3. Addressing Challenges and Ethical Considerations

While the principles of dignity and respect are crucial in healthcare, there are challenges and ethical considerations that need to be addressed. Some situations may involve a conflict between a healthcare professional's duty to promote a patient's well-being and the patient's autonomous decision-making.

In such cases, a delicate balance must be struck between respecting autonomy and ensuring the patient's best interests. This may involve exploring alternatives, engaging in open dialogue, and considering the potential consequences of the patient's choices. In extreme cases where a patient's decisions may lead to harm or pose a threat to others, limitations on autonomy may be required, but these should always be the last resort.

Cultural competence is essential in upholding dignity and respect, particularly in diverse healthcare settings. Health professionals must be aware of and sensitive to cultural norms and beliefs that may influence patients' health decisions. This includes understanding cultural differences related to illness perception, treatment preferences, and end-of-life care.

4. Case Study: Balancing Autonomy and Best Interests

Maria, a 53-year-old woman, has been diagnosed with stage 4 cancer. Despite aggressive treatment options available, Maria decides to refuse further treatment based on her personal values and religious beliefs. Her healthcare team respects her autonomy but is concerned about the potential impact of her decision on her overall well-being.

In this case, it is essential to engage in open and honest communication with Maria. The healthcare team should explore her reasons for refusing treatment, address any misconceptions, and offer alternative options that may align with her values, such as palliative care or pain management.

It is crucial to respect Maria's autonomy and support her decision while also ensuring that she fully understands the potential consequences. Continual monitoring and evaluation of her physical and emotional well-being are necessary, along with ongoing communication to address any potential changes in her treatment preferences.

By upholding the principles of dignity and respect, healthcare professionals can navigate complex ethical dilemmas and promote the overall well-being of their patients. Through open communication, cultural competence, and the provision of person-centered care, the healthcare system can truly honor and value the dignity of each individual.

Key Takeaways

- Dignity and respect are fundamental principles in health and wellness ethics, encompassing the inherent worth of each individual and their autonomous decision-making. - Person-centered care, respect for privacy, and confidentiality are essential in upholding dignity. - Respecting autonomy involves informed consent, involving patients in shared decision-making, and considering cultural and religious beliefs. - Balancing autonomy and best interests may require open dialogue, exploring alternatives, and considering potential consequences. - Cultural competence is crucial in respecting the diverse values and beliefs of patients.

Ethical Issues in Health and Wellness Promotion

Balancing Individual Autonomy and Public Health

In the realm of health and wellness promotion, ethical considerations often arise when trying to strike a balance between individual autonomy and public health. On one hand, individuals have the right to make their own decisions regarding their health and well-being, based on their personal values, beliefs, and preferences. On the other hand, public health initiatives aim to protect and improve the health of the population as a whole, which may sometimes require restricting certain individual freedoms. Finding the right balance between these two principles is crucial to promote both individual and societal well-being.

Understanding Individual Autonomy

Autonomy is a fundamental principle in bioethics that emphasizes a person's right to make decisions about their own lives, including their health and well-being. It recognizes that individuals have unique values, preferences, and goals, and should have the freedom to act according to their own choices, as long as those actions do not harm others. Respect for individual autonomy is considered a cornerstone of biomedical ethics and is enshrined in many ethical codes and documents, such as the principle of informed consent.

In the context of health and wellness promotion, individual autonomy means that individuals should have the right to make decisions about their own health behaviors, such as the food they eat, the exercise they engage in, or the medical treatments they undergo. It acknowledges that people have different beliefs, cultural backgrounds, and personal circumstances that influence their health-related choices. Respecting individual autonomy in these situations means allowing individuals to have agency over their own bodies and lives.

Promoting Public Health

While individual autonomy is important, public health also plays a crucial role in safeguarding the well-being of the overall population. Public health initiatives aim to prevent disease, promote healthy behaviors, and ensure access to essential healthcare services for everyone. These efforts often require implementing measures that may impact individual autonomy in order to protect the public's health.

For example, public health interventions such as vaccination mandates or restrictions on smoking in public places are implemented to reduce the spread of diseases and minimize harm to the broader community. These measures may restrict individual freedoms for the greater good of public health. However, it is essential to ensure that these interventions respect individual rights as much as possible and are based on scientific evidence and ethical principles.

Ethical Challenges

Balancing individual autonomy and public health can present several ethical challenges. One of the primary challenges is determining when and to what extent individual freedoms should be limited for the sake of public health. This requires considering factors such as the severity of the public health concern, the potential harms and benefits of public health interventions, and the likelihood of achieving the intended public health goals.

Another challenge is addressing potential conflicts between individual and societal values. People may have deeply held personal beliefs or cultural practices that conflict with public health recommendations. Respecting individual autonomy in these cases while also promoting public health can be complex and may require finding creative solutions that accommodate diverse perspectives.

Furthermore, ensuring that public health interventions do not disproportionately burden marginalized or vulnerable populations is a critical ethical consideration. It is important to recognize and address existing health inequities that may result from implementing public health measures, as certain groups may experience greater barriers to accessing healthcare or face disproportionate negative consequences.

Strategies for Balancing Autonomy and Public Health

To navigate the ethical challenges of balancing individual autonomy and public health, several strategies can be employed:

1. **Education and Communication:** Providing individuals with accurate and accessible information about the rationale behind public health measures can help them make informed decisions. Open and transparent communication can foster understanding and cooperation between individuals and public health authorities.

2. **Respecting Cultural Diversity:** Recognizing and respecting cultural diversity is crucial when developing and implementing public health interventions. Engaging with communities and understanding their unique values, beliefs, and practices can help tailor interventions that are sensitive to cultural differences and minimize conflicts with individual autonomy.

3. **Ethical Decision-making Frameworks:** Using ethical frameworks, such as the principles of beneficence, non-maleficence, and distributive justice, can guide decision-making when balancing individual autonomy and public health. These frameworks provide a systematic approach to evaluate the potential benefits and harms of interventions and consider the distribution of these benefits and harms among different populations.

4. **Collaboration and Partnership:** Engaging various stakeholders, including individuals, communities, healthcare professionals, and policymakers, in the decision-making process can foster collaboration and ensure that a wide range of perspectives are considered. This can lead to more ethically sound and socially acceptable public health measures.

5. **Evaluation and Adaptation:** Regularly evaluating the impact and outcomes of public health interventions is essential to ensure that they are achieving

their intended goals while minimizing unintended consequences. This evaluation process can help identify areas where ethical concerns need to be addressed and allow for adjustments to be made accordingly.

Case Study: COVID-19 Vaccination and Individual Autonomy

The COVID-19 pandemic has brought the issue of balancing individual autonomy and public health to the forefront. Vaccination campaigns have been implemented worldwide to control the spread of the virus and protect vulnerable populations. However, vaccine hesitancy and concerns about the infringement on personal freedoms have emerged as challenges.

In this case, public health authorities have a responsibility to provide accurate information about the safety and efficacy of vaccines while respecting individual autonomy and allowing individuals to make informed decisions. Utilizing education, communication, and collaboration, public health efforts can address the concerns and misconceptions surrounding vaccination and promote the greater good of public health.

However, in situations where individual decisions may pose a significant risk to public health, such as refusing vaccination in certain high-risk settings, policymakers may need to consider implementing specific measures to protect public health, while also striving to minimize harm to individual autonomy.

Conclusion

Balancing individual autonomy and public health in the context of health and wellness promotion is an ongoing ethical challenge. Respecting individual autonomy while considering the societal impact of health behaviors is crucial to ensure a just and sustainable future. By employing strategies such as education, respecting cultural diversity, ethical decision-making frameworks, collaboration, and evaluation, we can navigate these ethical dilemmas and promote both individual well-being and the health of the broader community. It is through this delicate balance that we can strive for a more equitable and just health and wellness future.

Nudging and Manipulation in Health Behavior Change

In the field of health and wellness, promoting behavior change is often a key objective to improve overall well-being and prevent diseases. However, influencing individuals to adopt healthier behaviors can be challenging. One approach that has gained attention in recent years is the use of nudges.

Understanding Nudges

Nudges can be defined as subtle changes in the environment that aim to influence people's behavior in a predictable way, without forbidding any options or significantly altering economic incentives. This concept was popularized by Richard Thaler and Cass Sunstein in their book "Nudge: Improving Decisions About Health, Wealth, and Happiness".

The underlying idea of nudges is that many of our choices are not the result of deliberate and rational decision-making processes. Instead, our behaviors are often shaped by cognitive biases, heuristics, and the context in which decisions are made. Nudges leverage these behavioral tendencies to encourage individuals to make healthier choices without restricting their freedom of choice.

Examples of Nudges in Health Behavior Change

Nudges can be implemented in various settings to promote health behavior change. Here are some examples:

+ **Healthy defaults:** By making healthy options the default choice, individuals are more likely to select them. For example, setting fruits and vegetables as the default options in school lunch programs can increase their consumption.

+ **Choice architecture:** Modifying the way choices are presented can influence decisions. Placing healthy snacks at eye level and less healthy options out of sight in cafeterias or supermarkets can encourage healthier choices.

+ **Social norms:** People tend to conform to what they believe is the norm. Communicating information about the prevalence of healthy behaviors can motivate individuals to adopt similar behaviors. For instance, displaying messages such as "8 out of 10 people wash their hands regularly" in public restrooms can increase handwashing compliance.

+ **Prompting reminders:** Providing timely reminders can promote healthier behaviors. Sending text messages reminding individuals to exercise, take medication, or attend health check-ups can reinforce positive habits.

+ **Gamification:** Applying game elements, such as rewards, challenges, or progress tracking, to health-related activities can make them more engaging and motivating. Fitness apps that reward users for meeting daily step goals or completing workout challenges are examples of gamification.

+ **Visual cues:** Using visual cues to highlight the health consequences of certain behaviors can influence decision-making. Graphic warning labels on cigarette packages or images depicting the detrimental effects of excessive sugar consumption on beverage containers are examples of visual cues.

Ethical Considerations

While nudges have the potential to be effective in promoting health behavior change, ethical concerns have been raised regarding their use. It is essential to ensure that nudges are transparent, respect individual autonomy, and do not manipulate or exploit vulnerable populations.

1. **Transparency:** Nudges should be openly communicated to individuals to maintain transparency and avoid deception. People should be aware that their choices are being influenced and understand the intention behind the nudge.

2. **Autonomy:** Nudges should not restrict people's freedom of choice or bypass their rational decision-making processes. Individuals should still have the option to opt-out or make alternative choices if they disagree with the nudged option.

3. **Equity and fairness:** Nudges should be implemented in a way that does not disproportionately target specific populations or reinforce existing social inequalities. Special attention should be given to ensuring that vulnerable populations are not exploited or manipulated.

Challenges and Limitations

Despite the potential benefits of nudges, they are not without challenges and limitations. It is crucial to recognize these limitations when considering their use in health behavior change interventions.

1. **Effectiveness:** Nudges may not work equally well for everyone. The effectiveness of nudges can vary depending on individual differences, cultural backgrounds, and personal motivations. It is essential to consider the diversity of the target population when designing nudges.

2. **Long-term sustainability:** Nudges may result in temporary behavior change, but sustaining these changes over the long term can be challenging. It is crucial to accompany nudges with comprehensive education, support systems, and policies that reinforce healthy behaviors.

3. **Potential backlash:** In some cases, individuals may react negatively to perceived paternalism or manipulation. Unintended consequences, such as reactance or resistance, can arise if individuals feel their autonomy is being

infringed upon. Careful consideration should be given to the implementation and communication of nudges to minimize potential backlash.

Conclusion

Nudging offers a promising approach to promoting health behavior change by leveraging cognitive biases and the contextual factors that influence our decisions. By carefully implementing nudges while considering ethical principles and being mindful of their limitations, we can harness their potential to create positive and sustainable changes in health and wellness behaviors. However, it is crucial to continue researching and evaluating the effectiveness and ethical implications of nudges to ensure their responsible and equitable application in practice.

Genetic Testing and Health Risk Communication

Genetic testing has become increasingly accessible in recent years, allowing individuals to gain insights into their genetic makeup and potential health risks. This section explores the ethical considerations surrounding genetic testing, as well as the importance of effective health risk communication in this context.

Understanding Genetic Testing

Genetic testing is the process of analyzing an individual's DNA to identify variations or alterations in genes associated with specific conditions or diseases. It can provide valuable information about an individual's risk of developing certain genetic disorders, as well as their potential response to certain medications.

There are different types of genetic tests, including diagnostic testing, predictive testing, carrier testing, and pharmacogenetic testing. Diagnostic testing is performed when a specific genetic disorder is suspected based on symptoms, while predictive testing is used to assess an individual's risk of developing a specific condition. Carrier testing is carried out to determine if an individual carries a genetic variant associated with a certain disorder, which can be crucial for family planning. Pharmacogenetic testing helps predict a person's response to certain medications based on their genetic profile.

Ethical Considerations

Genetic testing raises several ethical considerations that need to be carefully addressed. One of the primary concerns is the protection of an individual's privacy and confidentiality. Genetic information is highly personal, and there is a risk that it could be used to discriminate against individuals in various areas, such as employment and insurance coverage. Therefore, it is essential to ensure strict privacy measures and regulate the use and storage of genetic information.

Equity and access to genetic testing also pose ethical challenges. Genetic testing can be costly, making it inaccessible to individuals from socioeconomically disadvantaged backgrounds. This creates a potential disparity in access to crucial health information and the opportunity for timely interventions. Efforts should be made to reduce these disparities and ensure equitable access to genetic testing for all individuals.

Informed consent is another critical ethical principle in genetic testing. Individuals should be adequately informed about the benefits, limitations, and potential risks associated with genetic testing before undergoing the procedure.

This requires clear and thorough communication to ensure individuals can make informed decisions about their healthcare.

Furthermore, genetic counselors play a vital role in the ethical delivery of genetic testing services. They provide support and guidance to individuals and families in understanding the implications of their test results, disclosing potential risks, and exploring available options. Genetic counselors must adhere to ethical guidelines and maintain open and honest communication with patients.

Health Risk Communication

Effective health risk communication is crucial when conveying genetic test results to individuals. It involves delivering complex information in a clear and understandable manner to ensure individuals can comprehend the implications of their test results, make informed decisions, and take appropriate actions.

One of the key challenges in health risk communication is the accurate interpretation and understanding of probabilities. Genetic testing often provides probabilistic information about an individual's risk of developing a particular condition. However, accurately conveying these probabilities can be challenging, as individuals may have different levels of numeracy and may interpret probabilities differently. Healthcare providers and genetic counselors should use appropriate strategies, such as visual aids and plain language, to facilitate understanding.

Additionally, health risk communication should be culturally sensitive and tailored to the individual's preferences and values. Different cultures may have varying beliefs about genetic conditions, and it is crucial to respect and acknowledge these beliefs to ensure effective communication. It is also important to consider the emotional impact of the information being communicated and provide appropriate support and counseling throughout the process.

Furthermore, the media and online platforms play a significant role in shaping public perceptions and understanding of genetic testing. It is essential to promote accurate and balanced information, counter misinformation, and ensure that individuals have access to reliable resources for further education and support.

In conclusion, genetic testing and health risk communication are rapidly evolving areas with significant ethical implications. Protecting privacy, ensuring equity in access, obtaining informed consent, and delivering clear and culturally sensitive information are essential components of ethical genetic testing practices. Effective health risk communication is crucial for individuals to understand and act upon their genetic test results, ensuring they can make informed decisions about their health and well-being.

Media Influences on Body Image and Self-Worth

The media plays a powerful role in shaping societal norms and values, including those related to body image and self-worth. Through various outlets such as television, movies, magazines, and social media, the media constantly bombards individuals with images and messages about the ideal body and beauty standards. This relentless exposure can have significant implications for individuals' self-perception, self-esteem, and overall mental health. In this section, we will explore the ethical issues surrounding media influences on body image and self-worth, and discuss strategies for promoting a healthier and more inclusive media landscape.

The Impact of Media Portrayals on Body Image

The media frequently portrays narrow and unrealistic standards of beauty, often emphasizing thinness for women and muscularity for men. These portrayals can create a distorted perception of what constitutes a normal or desirable body, leading to body dissatisfaction and increased risk of developing eating disorders, such as anorexia nervosa or bulimia nervosa.

Research has shown that exposure to idealized images of physical attractiveness can negatively affect body image, particularly among vulnerable populations, such as adolescents and young adults. Studies have also highlighted the role of social comparison in this process, as individuals compare their own bodies to those depicted in the media, leading to feelings of inadequacy and dissatisfaction.

Furthermore, media messages can perpetuate stereotypes related to race, ethnicity, age, and gender, contributing to feelings of exclusion or marginalization among individuals whose bodies do not align with these ideals. For example, the underrepresentation of diverse body sizes, shapes, and skin colors in mainstream media can reinforce the notion that only a certain type of body is attractive or acceptable.

Ethics of Media Representation

The media has a responsibility to ethically represent and portray diverse body types and identities. Failure to do so can contribute to discrimination, body shaming, and the perpetuation of harmful stereotypes. Ethical considerations in media representation include:

1. Diversity and inclusivity: Media outlets should strive to reflect the diversity of the population in their content, including individuals of different races, ethnicities,

body sizes, shapes, and abilities. This can help to normalize a wider range of body types and reduce the negative impact of media on body image.

2.   Authenticity and truthfulness: The media should avoid editing or airbrushing images to an unrealistic extent, as this can create unattainable beauty standards and harm individuals' self-esteem.   Authentic representation can promote body acceptance and challenge unrealistic beauty ideals.

3. Responsible advertising practices: Advertisements should accurately portray the benefits and limitations of products and services and avoid using body shaming or negative messaging to sell products. Advertisers should also avoid promoting harmful weight loss or beauty enhancement techniques that endanger individuals' health.

4. Education and awareness: Media literacy programs should be implemented to help individuals critically evaluate and analyze media messages.   Education initiatives can empower individuals to challenge unrealistic beauty standards and make informed choices regarding their engagement with media content.

Promoting Positive Body Image

Promoting positive body image and self-worth requires collective efforts from individuals, media outlets, advertisers, and society as a whole. Here are some strategies that can help create a healthier media environment:

1. Media literacy education: Schools, community organizations, and healthcare professionals can implement media literacy programs to enhance critical thinking skills and promote awareness of media influence on body image. These programs can teach individuals to challenge unrealistic beauty ideals and engage with media content in a more mindful way.

2. Diverse and inclusive media representation: Media outlets should actively seek out and promote diverse voices and bodies in their content. This includes featuring individuals of different backgrounds, body types, and abilities, and avoiding harmful stereotypes or tokenism.   Increased representation can help broaden society's definition of beauty and promote acceptance.

3. Responsible advertising practices: Advertisers should adhere to ethical guidelines that promote body positivity and avoid perpetuating harmful standards. By utilizing models of various sizes, shapes, and backgrounds, advertisers can contribute to a more inclusive and diverse representation of beauty.

4. Encouraging positive social media use: Social media platforms can be both a source of harm and a platform for positive change. Encouraging users to follow diverse and body-positive accounts, as well as spreading messages of body acceptance, can promote a healthier social media environment.

5. Support for individuals with body image concerns: Healthcare professionals should be trained to recognize and address body image concerns in their patients. Providing appropriate support and resources can help individuals develop a more positive body image and build resilience against media influences.

In conclusion, media influences on body image and self-worth have significant ethical implications. The media should strive to represent a diverse range of bodies and identities authentically and responsibly. By promoting positive body image through media literacy education, diverse media representation, responsible advertising practices, and support for individuals with body image concerns, we can create a more equitable and inclusive media landscape that fosters self-acceptance and promotes well-being.

The Role of Industry in Health and Wellness Promotion

In the realm of health and wellness promotion, the role of industry is significant and influential. Various industries, including the pharmaceutical, food and beverage, fitness, and technology sectors, play a crucial role in shaping individuals' health behaviors and overall well-being. This section will explore the ethical considerations and potential challenges associated with the involvement of industry in health and wellness promotion, as well as the opportunities for collaboration and positive impact.

1. The Influence of Industry on Health Behavior:

The industry's influence on health behaviors is multifaceted, with both positive and negative implications. On one hand, industry initiatives can raise awareness about health issues, develop innovative products, and provide resources for individuals to make informed choices about their health. For example, the promotion of exercise equipment, health apps, and wearable technologies by the fitness and technology industries has encouraged individuals to be more physically active and monitor their health.

On the other hand, the influence of certain industries, such as the food and beverage industry, can contribute to unhealthy behaviors and the prevalence of chronic diseases. The marketing and advertising practices of the food and beverage industry, particularly towards children, have been criticized for promoting unhealthy food choices and contributing to the obesity epidemic.

2. Ethical Considerations:

The involvement of industry in health and wellness promotion raises several ethical considerations. One such consideration is the potential conflict of interest between profit-making goals and public health goals. The primary objective of any industry is to generate profits, which may not always align with the public health objective of improving individuals' well-being. This misalignment can lead to the prioritization of profit over health, as evidenced by the marketing of unhealthy products or the suppression of negative health information.

Furthermore, the industry's use of persuasive and manipulative marketing techniques raises concerns about autonomy and informed decision-making. Individuals may be influenced by misleading advertising or biased information, compromising their ability to make autonomous choices about their health. For example, the marketing of dietary supplements without sufficient scientific evidence may mislead individuals into investing in ineffective or potentially harmful products.

3. Regulation and Responsibility:

To address these ethical concerns, regulatory frameworks and industry responsibility are crucial. Governments and regulatory bodies play a significant role in monitoring and regulating industry practices to ensure they align with public health goals. For instance, labeling laws that require companies to provide accurate and clear information about the nutritional content of their products help individuals make informed choices.

Industry responsibility is another important aspect. Companies should adopt ethical marketing practices, provide accurate information, and prioritize the health and well-being of consumers. This involves avoiding deceptive advertising, engaging in transparent labeling, and promoting the responsible use of their products.

4. Collaboration for Positive Change:

Despite the challenges, industry involvement also presents opportunities for collaboration and positive change. Partnerships between the public health sector, nonprofit organizations, and industry can harness the resources and expertise of different sectors to promote health and wellness effectively. For example, collaborations between food companies and health organizations have resulted in the development of healthier product options and educational campaigns promoting healthy eating.

Additionally, industry can contribute to research and development efforts, leading to advancements in health technologies, therapies, and treatments. Collaborative research initiatives can help address health disparities, facilitate access to healthcare, and promote equity in health and wellness.

5. Case Study: Tobacco Industry and Public Health:

The tobacco industry serves as a critical example of the influence and ethical challenges associated with industry involvement in health and wellness. For many years, the tobacco industry fiercely resisted public health efforts to curb tobacco use by downplaying the health risks and targeting vulnerable populations. The industry's actions have been criticized for prioritizing profit over the health and well-being of individuals.

However, increased regulation, public awareness campaigns, and legal actions have led to significant progress in tobacco control. The implementation of comprehensive tobacco control policies, such as higher taxes, bans on advertising and marketing, and graphic warning labels, have contributed to a decline in smoking rates.

6. Conclusion:

In conclusion, the industry plays a significant role in health and wellness promotion, influencing individuals' behaviors and shaping public health outcomes. However, ethical considerations arise due to conflicts of interest and the potential for manipulative marketing practices. Regulatory frameworks, industry responsibility, and collaborations between different sectors are crucial to address these ethical challenges and harness the industry's potential for positive change. By working together, industry and public health stakeholders can strive towards a future where health and wellness promotion is guided by ethical principles and the well-being of individuals.

Social Determinants of Health and Wellness

Understanding Social Determinants of Health and Wellness

Definition and Key Concepts

In order to delve into the complexities of social determinants of health and wellness, it is crucial to establish a clear understanding of the concepts involved. This section will provide a comprehensive definition of social determinants, identify key elements, and explore the interconnectedness between social determinants and health outcomes.

Defining Social Determinants of Health and Wellness

Social determinants of health and wellness are the conditions in which people are born, grow, live, work, and age. These conditions are shaped by the distribution of money, power, and resources at global, national, and local levels. In essence, social determinants are the structural and societal factors that influence health and wellness outcomes, going beyond individual behaviors and biology. They encompass the broader socio-economic, political, cultural, and physical environments that individuals and communities are a part of.

At the core of social determinants is the acknowledgment that health and wellness disparities are not random or solely attributable to personal choices, genetics, or access to healthcare. Instead, they arise from systemic and structural factors that create and perpetuate inequalities. By focusing on social determinants, we can better understand how socio-economic, political, and environmental factors shape health and wellness outcomes.

Key Concepts of Social Determinants

To fully comprehend the impact of social determinants on health and wellness, it is important to grasp key concepts within this domain. The following concepts provide a framework for understanding the complex interactions between social determinants and health outcomes:

+ Social Justice: The ethical and moral belief in fair and equitable distribution of resources, opportunities, and privileges within a society. Social justice serves as a guiding principle for addressing health disparities rooted in social determinants.

+ Equity: The principle of fairness and justice. Achieving health equity means ensuring that everyone has an equal opportunity to attain their highest level of health and wellness, regardless of their social background or identity.

+ Intersectionality: The recognition and understanding of how different social identities intersect and interact to shape individuals' experiences and health outcomes. Intersectionality acknowledges that social determinants of health do not act in isolation but instead interact and compound to influence health disparities.

+ Social Gradient: The observation that health outcomes tend to worsen as one descends the social ladder. Higher socioeconomic status is associated with better health and wellness, while lower socioeconomic status is linked to increased health disparities.

+ Social Capital: The resources, relationships, and networks available within a community or society. Social capital has a significant impact on health and wellness as it affects access to resources, support systems, and opportunities for health promotion.

+ Health Inequalities: Disparities in health outcomes that are avoidable, unjust, and unfair. Health inequalities arise from social and economic conditions that result in differential access to resources and opportunities for health and wellness.

+ Life Course Perspective: The recognition that health and wellness outcomes are shaped by cumulative experiences and exposures over the course of an individual's life. This perspective emphasizes the importance of considering early life experiences, as well as the influence of social and environmental factors across the lifespan.

Interconnectedness between Social Determinants and Health Outcomes

Understanding the interconnectedness between social determinants and health outcomes is essential for addressing health disparities and promoting equitable health and wellness. Consider the example of socioeconomic status (SES). SES, which encompasses education, income, and occupation, significantly impacts health and wellness outcomes.

Individuals with higher SES generally have better access to healthcare, healthier living environments, and increased opportunities for health-promoting behaviors. On the other hand, individuals with lower SES often face barriers to healthcare, live in disadvantaged neighborhoods with limited resources, and experience higher levels of stress and adversity. These factors contribute to health disparities, with individuals of lower SES experiencing higher rates of chronic illnesses, poor mental health outcomes, and shorter life expectancy.

The interconnectedness between social determinants and health outcomes is complex and multifaceted. For instance, racial and ethnic disparities in health outcomes can be partially attributed to social determinants such as systemic racism, discrimination, and unequal distribution of resources. Similarly, gender inequities, including unequal access to education and employment opportunities, influence health outcomes for women and gender diverse individuals.

By recognizing the interconnectedness between social determinants and health outcomes, we can adopt an inclusive and holistic approach to promoting health and wellness. This entails addressing the root causes of health disparities and implementing multidimensional strategies that tackle the social, economic, and political factors that influence health outcomes.

Key Principles of Social Determinants of Health and Wellness

The study of social determinants of health and wellness is underpinned by several key principles that provide a framework for understanding and addressing health inequalities. These principles guide interventions, policies, and strategies aimed at promoting health equity and justice.

Principle of Redistribution

The principle of redistribution recognizes that resources and opportunities are distributed unevenly within societies, leading to health disparities. To promote health equity, it is essential to redistribute resources and power in a way that ensures everyone has equal access to the social determinants of health. This includes equitable distribution of income, education, employment, housing, and healthcare services.

Redistribution also involves addressing systemic and structural barriers that perpetuate health inequalities. By actively dismantling discriminatory policies, practices, and institutions, society can strive towards a more just and equitable distribution of resources and opportunities for health and wellness.

Principle of Proportionality

The principle of proportionality emphasizes that interventions and policies should be proportionate to the level of health disparity observed. It acknowledges that certain populations, such as those facing multiple intersecting forms of discrimination, may require greater attention and resources to address health disparities effectively.

Proportionate interventions prioritize those who are most affected by health inequalities and ensure that resources are allocated in a way that addresses the specific needs and contexts of marginalized populations. This principle highlights the importance of recognizing and addressing health disparities that are driven by social determinants, particularly for populations experiencing compounded forms of disadvantage.

Principle of Participation

The principle of participation emphasizes the inclusion and meaningful engagement of individuals and communities in decision-making processes that affect their health and well-being. Recognizing the expertise and lived experiences of marginalized communities is vital for designing and implementing interventions that address health disparities rooted in social determinants.

By involving individuals and communities in the planning, implementation, and evaluation of health promotion efforts, interventions can better reflect the diverse needs, priorities, and preferences of populations. This principle also empowers individuals and communities to advocate for policies and practices that promote health equity and challenge systemic injustices.

Principle of Accountability

The principle of accountability centers on holding governments, institutions, and systems responsible for addressing health inequalities and promoting health equity. It requires transparent and evidence-based decision-making processes, monitorin

Historical Overview of Social Determinants

To understand social determinants of health and wellness, it is important to examine their historical roots. The concept of social determinants emerged as a result of the recognition that health outcomes are shaped by factors beyond individual behaviors and genetics. This section will provide a historical overview of social determinants, tracing their origins, key developments, and significant milestones.

Origins of Social Determinants

The recognition that social factors have a profound impact on health can be traced back to the early 19th century. Social reformers, such as Edwin Chadwick and Rudolf Virchow, drew attention to the links between poverty, living conditions, and the health of populations. These early observers recognized that poor sanitation, inadequate housing, and limited access to healthcare were contributing to high disease rates and premature deaths among marginalized communities.

The social determinants framework gained further prominence in the mid-20th century with the publication of the Whitehall studies. These studies, led by Sir Michael Marmot, examined the health of British civil servants and found a clear social gradient in health outcomes. The higher the individual's socioeconomic position, the better their health. This research challenged the prevailing belief that health disparities were solely the result of individual choices and instead emphasized the role of social factors in shaping health.

Key Developments in Social Determinants

In the decades following the Whitehall studies, there have been several key developments in the understanding and application of the social determinants framework. These developments have contributed to a deeper understanding of the complex interactions between social factors and health outcomes.

One important development is the recognition of the importance of early life experiences. Research has shown that social disadvantage in early childhood can have long-lasting effects on health and well-being later in life. Adverse childhood experiences, such as abuse, neglect, and exposure to violence, have been linked to a higher risk of chronic diseases, mental health disorders, and even premature death. This understanding has led to an increased focus on early intervention and the importance of creating nurturing environments for children.

Another significant development is the recognition of the role of social support networks. Research has consistently shown that strong social connections and support systems are protective factors for health. On the other hand, social isolation and loneliness have been linked to a range of negative health outcomes, including cardiovascular disease, depression, and cognitive decline. This understanding has highlighted the importance of social networks in promoting health and wellness.

Milestones in Social Determinants

Over the years, there have been several milestones in the field of social determinants that have shaped both research and policy. One notable milestone is the publication of the World Health Organization's (WHO) Commission on Social Determinants of Health report in 2008. This report, titled "Closing the Gap in a Generation," highlighted the global nature of health inequities and called for action to address the social determinants that contribute to these inequities. The report emphasized the need for intersectoral collaboration, policy interventions, and political commitment to achieve health equity.

Another milestone is the inclusion of social determinants in the Sustainable Development Goals (SDGs) adopted by the United Nations in 2015. Goal 3 of the SDGs aims to ensure healthy lives and promote well-being for all at all ages. It recognizes that achieving this goal requires addressing the social determinants of health, such as poverty, education, gender inequality, and social exclusion. This global commitment has provided a framework for action to reduce health disparities and promote health equity worldwide.

Challenges and Future Directions

While significant progress has been made in understanding and addressing social determinants, challenges remain. One challenge is the complexity and interconnectedness of social determinants. These factors are not isolated but interact with each other in complex ways, making it challenging to identify and address individual determinants in isolation. A comprehensive and holistic approach is needed to tackle the multifaceted nature of social determinants.

Another challenge is the unequal distribution of power and resources that underlie social determinants. Addressing social determinants requires confronting systemic inequalities and structural barriers that perpetuate health disparities. This involves challenging social and economic systems that prioritize profit over people's health and well-being.

Moving forward, there is a need for continued research, policy development, and advocacy to address social determinants and promote health equity. This includes investing in upstream interventions that target the root causes of health inequities, promoting social and economic policies that reduce inequalities, and empowering marginalized communities to be agents of change.

Conclusion

The historical overview of social determinants reveals the evolution of our understanding of the complex and interconnected factors that shape health and wellness. From the early recognition of the link between poverty and disease to the current focus on structural determinants of health, the field has come a long way.

As we continue to explore and address social determinants, it is crucial to remember that health is not just an individual responsibility but a collective endeavor. By addressing social determinants, promoting justice, and striving for a more equitable and just world, we can create the conditions necessary for all individuals to live healthy and fulfilling lives.

Intersectionality and Health Disparities

In our exploration of the social determinants of health and wellness, we must consider the concept of intersectionality. Intersectionality refers to the interconnected nature of social categories such as race, gender, class, and sexuality, and how they overlap and intersect to create unique experiences and systems of privilege and marginalization. Recognizing intersectionality is essential when addressing health disparities, as it allows us to understand how multiple dimensions of identity can contribute to differential access to resources, opportunities, and healthcare.

The Concept of Intersectionality

Intersectionality was first introduced by legal scholar Kimberlé Crenshaw in 1989 to describe the experiences of Black women who faced both racial and gender discrimination but were often overlooked by feminist and antiracist movements. It recognizes that individuals do not experience oppression or privilege solely based on a single identity but rather through the combination of multiple identities.

For example, a black woman may face challenges that are distinct from those faced by a white woman or a black man due to the intersection of race and gender. This intersectional lens helps to uncover the complexities of discrimination and oppression that individuals with multiple marginalized identities face and to address the unique health disparities that result from these intersections.

Intersectionality and Health Disparities

Health disparities refer to differences in health outcomes or healthcare access between different groups of people. These disparities are not random but are rooted in social inequities related to race, ethnicity, socioeconomic status, gender, and other intersecting identities. Understanding and addressing the role of intersectionality in health disparities is essential to promote equitable and just health and wellness.

For example, studies have consistently shown that Black women in the United States have higher rates of maternal mortality compared to white women. This disparity cannot be solely attributed to race or gender; instead, it is the result of the intersection of race and gender, along with other factors such as socioeconomic status, access to healthcare, and systemic racism within the healthcare system.

Addressing Intersectionality in Health Interventions

To effectively address health disparities, it is crucial to incorporate an intersectional approach into health interventions and policies. Here are some key considerations:

1. **Holistic understanding of diversity:** Recognize that individuals embody multiple identities and that health disparities result from the interplay of these identities. This requires collecting disaggregated data to capture the experiences and needs of different subpopulations.

2. **Collaboration and inclusion:** Involve diverse stakeholders, including community members, marginalized groups, and interdisciplinary professionals, in decision-making processes related to health interventions. This ensures that interventions are sensitive to the unique needs and contexts of different intersecting identities.

3. **Tailored interventions:** Develop interventions that are sensitive to the complex realities of individuals' lives, including the intersecting dimensions of their identity. This may involve addressing social determinants of health such as education, employment, housing, and access to affordable healthcare.

4. **Intersectional research:** Conduct research that explicitly explores the intersections of different identities and their impact on health outcomes. This research should employ qualitative and quantitative methods to capture the multifaceted nature of health disparities and inform evidence-based interventions.

5. Health workforce diversity: Promote diversity in the healthcare workforce by encouraging the recruitment and retention of individuals from marginalized groups. This helps to ensure cultural competence, reduce biases, and improve access to healthcare services for diverse populations.

Case Study: Intersectionality and Mental Health

To illustrate the impact of intersectionality on health disparities, let's consider the mental health experiences of LGBTQ+ youth. LGBTQ+ individuals often face discrimination and stigma based on their sexual orientation or gender identity. However, when we consider the intersectionality of race, socioeconomic status, and other factors, we find that mental health disparities are even more pronounced.

For example, a study found that compared to their white LGBTQ+ peers, LGBTQ+ youth of color reported higher levels of depression, anxiety, and suicidal ideation. Intersectional factors such as racism and socioeconomic inequalities contribute to these disparities, highlighting the need for interventions that address both sexual orientation and race-related stressors.

Conclusion

Intersectionality provides a framework for understanding the complexities of health disparities by acknowledging the interplay of multiple identities and social categories. By recognizing the ways in which different axes of oppression intersect, we can develop more comprehensive and effective strategies to promote health and wellness equity. It challenges us to address health disparities at their root, taking into account the broader social, economic, and political structures that contribute to differential health outcomes. Embracing intersectionality can lead us towards a more just and equitable future for health and wellness.

Economic Inequalities and Health

Poverty and Health Disparities

In this section, we will explore the relationship between poverty and health disparities. Poverty is a significant social determinant of health that affects individuals and communities worldwide. It plays a crucial role in shaping health outcomes, access to healthcare services, and overall well-being. Understanding the dynamics of poverty and its impact on health is essential for developing effective interventions and promoting health equity.

The Link Between Poverty and Health

Poverty is often characterized by inadequate income, lack of access to basic amenities, and limited resources. These conditions can have a profound impact on health and well-being. People living in poverty face multiple challenges that increase their susceptibility to illness and disease.

One key factor is the limited access to healthcare services. Individuals in poverty are less likely to have health insurance coverage and regular access to primary care providers. As a result, they may delay seeking care, leading to the progression of illnesses and poorer health outcomes. Limited financial resources may also restrict their ability to afford essential medications and treatments.

Moreover, poverty is associated with a higher prevalence of unhealthy lifestyle behaviors such as smoking, poor nutrition, and sedentary lifestyles. These behaviors contribute to the development of chronic diseases such as cardiovascular disease, diabetes, and obesity.

Furthermore, poverty often intersects with other social determinants of health, such as education, employment, and housing. Limited educational opportunities and low educational attainment can perpetuate the cycle of poverty and affect health outcomes. Unemployment and unstable employment can further exacerbate financial stress and limit access to healthcare services. Inadequate and unsafe housing conditions pose additional health risks, including increased exposure to environmental pollutants and infectious diseases.

Health Disparities and Poverty

Health disparities refer to differences in health outcomes and access to healthcare services across different population groups. Poverty is a major driver of health disparities, as it intersects with other social determinants of health. Certain population groups, such as racial and ethnic minorities, are disproportionately affected by poverty and experience higher rates of health disparities.

For example, studies have consistently shown that racial and ethnic minority groups experience higher rates of poverty compared to the majority population. As a result, they face increased barriers to healthcare access, limited healthcare resources, and higher incidence rates of chronic diseases. This contributes to significant health disparities, including higher mortality rates and poorer health outcomes.

Additionally, poverty-related health disparities are not limited to low-income countries. In high-income countries, such as the United States, income inequality and poverty disproportionately affect marginalized populations. For instance, individuals living in inner-city neighborhoods with high poverty rates may face limited access to quality healthcare, leading to higher rates of preventable diseases and premature mortality.

Addressing Poverty and Health Disparities

Addressing poverty and health disparities requires a multi-faceted approach that addresses the underlying social and economic determinants of health. Efforts should focus on promoting equitable access to healthcare services, addressing systemic barriers, and improving living conditions.

One key intervention is the expansion of healthcare coverage, including initiatives for universal healthcare. By ensuring that all individuals have access to essential healthcare services regardless of their financial status, we can reduce disparities and improve health outcomes.

Furthermore, implementing policies that uplift socioeconomically disadvantaged populations is crucial. This may include increasing the minimum wage, providing affordable housing, and expanding educational opportunities. By improving living conditions and addressing structural factors contributing to poverty, we can create a more equitable society where health disparities are minimized.

Community-based interventions and outreach programs are also essential in addressing poverty and health disparities. These initiatives provide education, resources, and support to individuals and communities living in poverty. They empower individuals to make healthier choices, access preventive care, and navigate the healthcare system.

Lastly, research and data collection are vital for identifying and monitoring health disparities related to poverty. By understanding the specific health needs of different populations and evaluating the effectiveness of interventions, we can develop targeted strategies for reducing disparities.

Case Study: The Health Impact of Poverty in Sub-Saharan Africa

Sub-Saharan Africa is a region greatly affected by poverty and health disparities. Economic inequalities, political instability, and limited resources contribute to high poverty rates and poor health outcomes. The health impact of poverty in this region is evident in several key areas.

Firstly, there is a high prevalence of infectious diseases in Sub-Saharan Africa, such as malaria, tuberculosis, and HIV/AIDS. Poverty creates conditions that facilitate the spread of these diseases, including inadequate housing, lack of access to clean water and sanitation, and limited healthcare infrastructure. As a result, the burden of these diseases is disproportionately borne by individuals and communities living in poverty.

Secondly, maternal and child health is significantly impacted by poverty. Limited access to quality prenatal care, nutrition, and adequate healthcare facilities contributes to higher rates of maternal and infant mortality. The socioeconomic determinants of health, such as education and income, influence access to essential maternal and child health services.

Lastly, chronic diseases, such as cardiovascular disease and diabetes, are on the rise in Sub-Saharan Africa. While infectious diseases still pose a significant health burden, the changing disease landscape calls for a comprehensive response to combat both communicable and non-communicable diseases. Poverty, with its influence on lifestyle factors and limited access to healthcare, plays a critical role in the increasing prevalence of chronic diseases.

Addressing the health impact of poverty in Sub-Saharan Africa requires a comprehensive approach. This includes investing in healthcare infrastructure, increasing access to essential medicines, strengthening public health systems, and promoting socioeconomic development. Community engagement, education, and empowerment are central to overcoming the challenges posed by poverty and achieving health equity.

Conclusion

In this section, we explored the link between poverty and health disparities. Poverty, as a social determinant of health, has a significant impact on individuals and communities worldwide. It affects access to healthcare services, unhealthy lifestyle behaviors, and the overall well-being of individuals living in poverty.

Health disparities related to poverty exist across different population groups and countries. Racial and ethnic minorities, as well as marginalized populations, are disproportionately affected by poverty and experience higher rates of health disparities.

Addressing poverty and health disparities requires a multi-faceted approach that targets the underlying social and economic determinants of health. This includes expanding healthcare coverage, implementing policies that uplift socioeconomically disadvantaged populations, and promoting community-based interventions and outreach programs.

Furthermore, research and data collection are vital for identifying and monitoring health disparities related to poverty. By understanding the specific health needs of different populations and evaluating the effectiveness of interventions, we can develop targeted strategies for reducing disparities and promoting health equity.

It is crucial that we continue to strive for a more equitable and just health and wellness future, where poverty and its impact on health are addressed comprehensively. By prioritizing the reduction of poverty and tackling its determinants, we can work towards creating a world where everyone has an equal opportunity to achieve optimal health and well-being.

Access to Healthcare and Health Insurance

Access to healthcare and health insurance is a critical issue in the field of health and wellness. In this section, we will explore the importance of access to healthcare services, the challenges individuals face in obtaining access, and the ethical considerations surrounding the availability and affordability of health insurance.

Understanding the Importance of Access to Healthcare

Access to healthcare is a fundamental right that ensures individuals can receive timely and appropriate medical care without facing barriers. It is essential for promoting and maintaining the health and wellbeing of individuals and communities.

Without access to healthcare services, individuals may delay or forgo necessary medical treatments, leading to worsened health outcomes and increased healthcare costs in the long run. Accessible healthcare services help prevent and manage chronic diseases, improve overall health status, and reduce healthcare disparities.

Challenges in Accessing Healthcare

Despite the importance of access to healthcare, many individuals face significant challenges in obtaining the care they need. These challenges can be attributed to various factors, including economic, geographical, social, and cultural barriers.

+ **Economic Barriers:** High healthcare costs, lack of health insurance coverage, and limited financial resources can create significant barriers to accessing healthcare services. The cost of medical treatments, prescription medications, and preventive services can be prohibitively expensive for many individuals and families.

+ **Geographical Barriers:** Accessibility to healthcare services can be limited in rural or remote areas where healthcare facilities and providers are scarce.

Limited transportation options and long travel distances to healthcare centers often pose challenges for individuals residing in these areas.

+ **Social and Cultural Barriers:** Language barriers, lack of cultural competency among healthcare providers, and discriminatory practices can hinder access to healthcare for marginalized populations, including minority communities and immigrants. Socioeconomic factors, such as low health literacy levels and high levels of mistrust in the healthcare system, can also contribute to barriers in accessing care.

The Role of Health Insurance

Health insurance plays a crucial role in ensuring access to healthcare services. It provides financial protection by covering a portion of the costs associated with medical treatments, medications, and preventive services. By spreading the financial risk across a pool of insured individuals, health insurance helps reduce the burden of healthcare expenses on individuals and families.

Types of Health Insurance

There are various types of health insurance coverage models, each with its own characteristics and requirements. Some common types of health insurance include:

+ **Employer-Sponsored Health Insurance:** Many individuals receive health insurance coverage through their employers. Employer-sponsored health insurance plans are typically offered as part of an employee benefits package, with the employer and employee sharing the cost of premiums.

+ **Government-Sponsored Health Insurance:** Governments at various levels provide health insurance coverage to eligible individuals and groups. Examples include Medicare and Medicaid in the United States, which provide coverage for elderly and low-income individuals, respectively. Other countries have their own government-sponsored health insurance programs.

+ **Individual Health Insurance:** Individuals who do not have access to employer-sponsored or government-sponsored health insurance can purchase individual health insurance plans directly from insurance providers. These plans typically require individuals to pay a monthly premium in exchange for coverage.

+ **Community-Based Health Insurance:** In some communities or regions, community-based health insurance programs exist to provide coverage for individuals who may not qualify for other types of insurance. These programs are often created by local organizations or nonprofits to address specific healthcare needs within their communities.

Ethical Considerations in Health Insurance

The availability and affordability of health insurance raise important ethical considerations. Here are some key ethical issues related to health insurance:

+ **Equity and Fairness:** Health insurance should be designed in a way that promotes equity and fairness, ensuring that everyone has an equal opportunity to access healthcare services. This requires addressing disparities in access and affordability, particularly for marginalized and underserved populations.

+ **Solidarity and Shared Responsibility:** Health insurance operates on the principle of solidarity and shared responsibility, where everyone contributes to the collective pool of funds to support healthcare services for all. This principle ensures that the healthcare costs are distributed fairly among the insured population.

+ **Risk Pooling and Pre-existing Conditions:** Health insurance should promote broad risk pooling to minimize the impact of health risks on individuals. This means that insurance providers should not discriminate against individuals with pre-existing conditions or charge them higher premiums based on their health status.

+ **Transparency and Informed Choice:** Health insurance providers should ensure transparency in their policies, coverage limits, and costs. Individuals should have access to clear and understandable information to make informed choices about their insurance coverage.

Addressing Access to Healthcare and Health Insurance

To improve access to healthcare and health insurance, a multidimensional approach is required. Here are some strategies that can help address the challenges individuals face in accessing healthcare:

+ **Universal Health Coverage:** Governments can work towards implementing universal health coverage, where everyone in a country or region is guaranteed access to essential healthcare services without experiencing financial hardship. This can be achieved through a combination of government-sponsored programs, employer-based coverage, and private insurance options.

+ **Healthcare System Strengthening:** Healthcare systems need to be strengthened to ensure an adequate supply of healthcare providers, particularly in underserved areas. Training and deploying healthcare professionals to areas with limited access can help improve geographical barriers to healthcare.

+ **Reducing Healthcare Costs:** Measures should be taken to reduce healthcare costs and ensure affordability for individuals. This can include price regulation for medical treatments and prescription medications, negotiating lower prices with healthcare providers, and promoting the use of cost-effective healthcare practices.

+ **Health Education and Awareness:** Health literacy programs can help individuals understand their health insurance options, rights, and benefits. By improving health literacy, individuals can make informed choices and navigate the healthcare system more effectively.

+ **Cultural Competence and Eliminating Bias:** Healthcare providers should receive training in cultural competence to ensure that they can provide equitable and culturally sensitive care to diverse populations. Eliminating bias and discrimination in healthcare delivery is crucial for promoting access and improving health outcomes.

Example

Consider the case of Maria, a low-income individual who does not have health insurance. Maria has a chronic health condition that requires regular medical appointments and prescription medications. However, due to the high cost of healthcare services, she has been unable to access the necessary care and has experienced worsening health outcomes.

To address Maria's challenges in accessing healthcare, policymakers can consider implementing a government-sponsored program that provides affordable health insurance to low-income individuals. This program can offer subsidies to cover premiums and out-of-pocket costs, ensuring that individuals like Maria can afford the medical care they need. Additionally, efforts can be made to increase the availability of healthcare providers in Maria's community, minimizing the geographical barriers she faces.

By addressing the economic and geographical barriers that hinder individuals' access to healthcare, policymakers can ensure that everyone has equal opportunities to receive the necessary medical care, regardless of their socioeconomic status or location.

Key Takeaways

Access to healthcare and health insurance is vital for promoting and maintaining the health and wellbeing of individuals and communities. However, various challenges, including economic, geographical, social, and cultural barriers, can hinder access to healthcare services.

Health insurance plays a crucial role in ensuring access to healthcare by providing financial protection and spreading the financial risk across a pool of insured individuals. Ethical considerations in health insurance include equity, fairness, solidarity, shared responsibility, and transparency.

Addressing access to healthcare and health insurance requires a multidimensional approach, including universal health coverage, strengthening healthcare systems, reducing healthcare costs, improving health literacy, and promoting cultural competence. These strategies can help promote equitable access to healthcare services and improve health outcomes for all individuals.

Socioeconomic Status and Health Outcomes

Socioeconomic status (SES) plays a crucial role in determining individual and population health outcomes. It refers to an individual's or a group's position within a social hierarchy, influenced by factors such as income, education, occupation, and wealth. Numerous studies have shown a strong association between socioeconomic status and health, with lower SES individuals experiencing higher rates of illness, disability, and premature death. In this section, we will explore the complex relationship between socioeconomic status and health outcomes, as well as the underlying mechanisms and ethical implications involved.

The Link between Socioeconomic Status and Health

First, let's examine the evidence supporting the link between socioeconomic status and health outcomes. Studies consistently demonstrate that higher socioeconomic status is associated with better health and a lower risk of chronic diseases. For instance, individuals with higher income and education levels tend to have longer life expectancies, lower rates of cardiovascular disease, and lower prevalence of obesity.

On the other hand, individuals from lower socioeconomic backgrounds often face disadvantaged living and working conditions, limited access to healthcare, and increased exposure to environmental hazards. These factors contribute to a higher burden of diseases such as diabetes, respiratory illnesses, and mental health disorders among individuals with lower SES.

Pathways and Mechanisms

The relationship between socioeconomic status and health outcomes can be understood through various pathways and mechanisms. These include:

1. **Material resources:** Higher SES individuals have access to better healthcare services, nutritious food, safer living conditions, and other essential resources that promote health and well-being. Conversely, individuals with lower SES may struggle to afford basic necessities or face barriers to accessing healthcare, leading to poorer health outcomes.

2. **Psychosocial factors:** Socioeconomic disadvantage is associated with chronic stress, social exclusion, and limited social support. These psychosocial factors can negatively impact mental health and physiological processes, increasing the risk of conditions like depression, anxiety, and cardiovascular diseases.

3. **Health behaviors:** SES influences health-related behaviors such as smoking, physical activity, and dietary choices. Higher SES individuals generally have greater resources and knowledge to engage in healthier behaviors, while lower SES individuals may face barriers to adopting and maintaining healthy lifestyles.

4. **Early-life conditions:** Adverse socioeconomic circumstances during early childhood, such as poverty, inadequate nutrition, and exposure to stress, can have long-lasting effects on health outcomes. These early-life conditions may influence biological development, gene expression, and the risk of chronic diseases later in life.

5. **Healthcare access and utilization:** Individuals with higher SES are more likely to have health insurance coverage, regular access to healthcare providers, and utilize preventive services. In contrast, individuals with lower

SES may face financial barriers, inadequate health insurance, and limited access to quality healthcare, leading to delayed or suboptimal healthcare utilization.

Ethical Considerations

The relationship between socioeconomic status and health outcomes raises important ethical considerations. It highlights the fundamental role of social justice and equitable distribution of resources in promoting population health. Some key ethical considerations include:

1. **Health inequalities:** Health disparities resulting from socioeconomic disadvantage represent a form of social injustice. Addressing these inequalities requires a comprehensive approach that includes policy changes, social programs, and efforts to reduce the socioeconomic gradient in health outcomes.

2. **Equitable access to healthcare:** Ensuring equitable access to healthcare is crucial in reducing health disparities related to socioeconomic status. Efforts should focus on improving healthcare infrastructure in underserved areas, providing financial assistance for low-income individuals, and implementing policies that address barriers to healthcare access.

3. **Social determinants of health:** Recognizing and addressing the social determinants of health, such as poverty, education, and employment, is essential in improving health outcomes for disadvantaged populations. Policies that promote social and economic opportunities can help break the cycle of poverty and improve overall health.

4. **Advocacy and empowerment:** Ethical considerations also emphasize the importance of advocacy and empowerment for marginalized populations. This involves amplifying the voices of individuals living in lower socioeconomic conditions, involving them in decision-making processes, and addressing their unique healthcare needs.

5. **Research and data collection:** Collecting and analyzing data on the relationship between socioeconomic status and health outcomes is crucial for evidence-based policymaking. Research should focus on understanding the underlying factors contributing to health disparities, evaluating the effectiveness of interventions, and monitoring progress towards achieving health equity.

Case Study: Socioeconomic Status and Diabetes

To illustrate the impact of socioeconomic status on health outcomes, let's consider the case of diabetes. Studies have consistently shown that individuals with lower SES are at a higher risk of developing diabetes and experience poorer diabetes management outcomes compared to their higher SES counterparts.

Various factors contribute to these disparities. Limited access to nutritious food in low-income neighborhoods can result in a higher prevalence of unhealthy diets, which is a major risk factor for diabetes. Additionally, individuals with lower SES may face challenges in affording medications, regular blood glucose monitoring, and accessing diabetes education programs, leading to poorer disease management.

Addressing these disparities requires a comprehensive approach. It involves implementing policies that promote healthy food environments in underserved communities, expanding access to affordable healthcare and diabetes management services, and advocating for income equality to address the underlying socioeconomic factors contributing to diabetes disparities.

Conclusion

Socioeconomic status significantly influences health outcomes, with lower SES individuals experiencing higher rates of illness, disability, and premature death. Understanding the complex relationship between socioeconomic status and health can guide efforts to promote health equity and address health disparities. By addressing the social determinants of health, advocating for equitable access to healthcare, and implementing policies that promote social justice, we can work towards a more just and equitable health and wellness future for all individuals, regardless of their socioeconomic status.

Global Poverty and Health Disparities

In this section, we will explore the intersection between global poverty and health disparities. Poverty is a complex social issue that affects individuals, families, and entire communities around the world. It impacts various aspects of people's lives, including their access to healthcare and overall health outcomes. We will examine the causes and consequences of global poverty, discuss its link to health disparities, and explore potential solutions to address this pressing issue.

Understanding Global Poverty

Global poverty refers to the state of extreme deprivation in which individuals lack basic necessities, such as clean water, adequate food, shelter, and access to healthcare. It is a multidimensional issue influenced by various interconnected factors, including economic, political, social, and environmental factors. Poverty is often concentrated in developing countries, where resources and opportunities are limited. However, poverty also exists in developed nations, albeit to a lesser extent.

Causes of Global Poverty

Several factors contribute to the persistence of global poverty. Economic factors, such as unequal distribution of wealth, limited job opportunities, and low wages, are significant contributors. Political factors, including corruption, lack of good governance, and political instability, can perpetuate poverty by obstructing economic development and social progress. Social factors like discrimination, social exclusion, and lack of access to education and healthcare further exacerbate poverty. Additionally, environmental factors such as climate change, natural disasters, and resource depletion disproportionately impact vulnerable populations, trapping them in a cycle of poverty.

Health Disparities and Poverty

Poverty and health disparities are intimately linked, as poverty hinders individuals' access to healthcare and exacerbates health issues. Limited financial resources and lack of health insurance make it difficult for individuals living in poverty to seek timely and appropriate medical care. Consequently, they often face delays in diagnosis, inadequate treatment, and higher rates of preventable diseases. Poverty also intersects with other social determinants of health, such as education, housing, and nutrition, further compounding health disparities.

Global Health Inequalities

Global health inequalities refer to the unequal distribution of health resources and opportunities across countries and populations. Developing countries, particularly those with high poverty rates, experience a disproportionately high burden of disease and lack access to basic healthcare services. Inadequate infrastructure, shortage of healthcare professionals, and limited resources contribute to poor health outcomes in these regions. As a result, individuals living in poverty are more likely to suffer from infectious diseases, malnutrition, maternal and child health issues, and other preventable conditions.

Addressing Global Poverty and Health Disparities

Addressing global poverty and health disparities requires a multifaceted approach involving governments, international organizations, healthcare professionals, and civil society. Here are some strategies to consider:

1. **Economic Development:** Promoting sustainable economic growth and reducing income inequality are crucial in combating global poverty. This includes investing in education, infrastructure, and job creation to enhance livelihood opportunities for individuals and communities.

2. **Universal Healthcare Coverage:** Implementing universal healthcare systems can ensure that everyone, including those living in poverty, has access to essential healthcare services without facing financial hardship. Governments should prioritize expanding healthcare infrastructure, providing affordable medicines, and training healthcare professionals in underserved areas.

3. **Education and Empowerment:** Providing quality education and vocational training equips individuals with the skills and knowledge necessary to escape poverty. Education also plays a vital role in promoting health literacy, enabling individuals to make informed decisions about their health and well-being.

4. **Social Protection Programs:** Establishing social safety nets, such as cash transfer programs and social insurance schemes, can protect vulnerable populations from the adverse effects of poverty. These programs provide financial support, promote access to healthcare, and help individuals break the cycle of poverty.

5. Global Partnerships: Collaboration between governments, international organizations, and civil society is essential to address global poverty and health disparities. Cooperation should focus on resource sharing, capacity building, and knowledge exchange to empower communities, strengthen healthcare systems, and promote equity in access to healthcare.

Case Study: The Role of Non-Governmental Organizations (NGOs)

Non-governmental organizations (NGOs) play a crucial role in addressing global poverty and health disparities. These organizations work on the ground, providing direct support to marginalized communities and advocating for policy changes. One noteworthy example is the Bill & Melinda Gates Foundation, which focuses on improving global health and reducing poverty. The foundation invests in research, implements effective interventions, and supports healthcare systems in developing countries.

Conclusion

Global poverty and health disparities are complex and interconnected challenges that require comprehensive and coordinated efforts to address. By understanding the causes and consequences of global poverty, implementing sustainable development strategies, promoting universal healthcare coverage, and fostering global partnerships, we can work towards a more equitable and just future for all, where everyone has access to the resources and opportunities necessary to lead a healthy and fulfilling life.

Note: It is essential to acknowledge that addressing global poverty and health disparities is an ongoing process, and there is no one-size-fits-all solution. Solutions should be tailored to the specific needs and contexts of different regions and populations. This requires continuous research, evaluation, and adaptation of strategies to ensure their effectiveness and sustainability.

Exercise: Conduct a case study on a specific country or region experiencing high poverty rates and health disparities. Analyze the factors contributing to these issues and propose a comprehensive intervention strategy that addresses economic, social, and healthcare dimensions. Consider the role of various stakeholders, including government, NGOs, and international organizations, in implementing and sustaining the intervention.

Racial and Ethnic Inequalities in Health

Historical Context of Racial and Ethnic Health Disparities

The historical context of racial and ethnic health disparities encompasses a range of factors that have contributed to the unequal distribution of health outcomes among different racial and ethnic groups. Understanding this context is essential for addressing and eliminating these disparities. In this section, we will explore the historical events, policies, and social determinants that have shaped the health status of marginalized communities.

Colonialism and Slavery

The origins of racial and ethnic health disparities can be traced back to the era of colonialism and slavery. During the transatlantic slave trade, Africans were forcibly brought to the Americas as slaves, enduring unimaginable conditions during the journey and throughout their lives. These enslaved individuals experienced severe physical and psychological trauma, lack of access to healthcare, and inadequate living conditions, leading to poor health outcomes for themselves and future generations.

In addition, colonial powers exploited the resources of indigenous populations, leading to displacement, loss of land, and exposure to new diseases. Indigenous communities faced devastating health consequences due to forced displacement, cultural disruption, and loss of traditional practices.

Segregation and Jim Crow Laws

Following the abolition of slavery in the United States, racial discrimination persisted through the implementation of segregation laws and policies, known as Jim Crow laws. These laws enforced racial separation and denied African Americans basic rights and opportunities, including access to quality healthcare, education, housing, and employment.

Segregation resulted in the creation of under-resourced and overcrowded neighborhoods, referred to as "ghettos" or "inner cities," predominantly inhabited by racial and ethnic minority populations. These neighborhoods lacked essential resources, including access to healthcare facilities, healthy food options, and green spaces. The living conditions in these areas contributed to the development of chronic diseases, such as heart disease and diabetes, and limited opportunities for overall health and wellness.

Medical Experimentation and Biases

Historically, marginalized communities have been subjected to unethical medical experimentation and biases in healthcare practices. One notorious example is the Tuskegee Syphilis Study, conducted between 1932 and 1972 in the United States. In this study, African American men with syphilis were intentionally left untreated, even after the discovery of a cure, leading to severe health complications and loss of life.

These unethical practices and biases have resulted in mistrust and hesitancy towards healthcare systems within racial and ethnic minority populations. This lack of trust has made it challenging to address health issues effectively, resulting in delayed healthcare-seeking behaviors, poorer health outcomes, and ongoing disparities.

Structural Racism

Structural racism refers to the ways in which policies, institutions, and practices systematically advantage certain racial and ethnic groups while disadvantaging others. It operates within the social, economic, and political structures of society, perpetuating racial and ethnic health disparities.

For instance, redlining, a discriminatory housing policy prevalent in the United States, systematically denied mortgage loans and insurance to individuals in predominantly minority neighborhoods. This led to limited access to quality housing, educational opportunities, healthcare resources, and safe environments, contributing to adverse health outcomes.

Additionally, institutional racism within healthcare systems, such as discriminatory practices in the allocation of resources and access to healthcare services, has perpetuated racial and ethnic health disparities. This includes disparities in preventive care, treatment options, and health outcomes for diseases such as cancer, cardiovascular disease, and maternal mortality.

Social Determinants of Health

Social determinants of health, including education, income, occupation, housing, and access to healthcare, significantly influence health outcomes. These determinants are shaped by historical and ongoing systemic factors, including racism, discrimination, and socio-economic disparities.

For example, individuals from racial and ethnic minority groups are more likely to experience poverty, limited educational opportunities, and unstable employment, which can have detrimental effects on health. The lack of quality education, lower income levels, and limited job prospects contribute to higher rates of chronic diseases, mental health disorders, and overall poorer health outcomes.

Additionally, racial and ethnic minority populations often face limited access to healthcare services, including primary care, preventive care, and specialized care, exacerbating health disparities.

Addressing Racial and Ethnic Health Disparities

Addressing racial and ethnic health disparities requires a comprehensive approach that tackles both the root causes and the immediate effects. Policies and interventions should aim to eliminate structural racism, improve social determinants of health, and ensure equitable access to quality healthcare for all individuals.

Some strategies to address racial and ethnic health disparities include:

+ Implementing policies that address the social determinants of health, such as providing affordable housing, quality education, and employment opportunities.

+ Promoting diversity and inclusivity in the healthcare workforce to improve cultural competence and reduce biases in healthcare delivery.

+ Engaging communities in decision-making processes to ensure their needs and perspectives are taken into account when designing health programs and policies.

+ Conducting research that focuses on understanding the underlying causes of racial and ethnic health disparities and developing evidence-based interventions to eliminate them.

+ Increasing access to healthcare services in underserved areas by expanding healthcare infrastructure, improving transportation systems, and implementing telemedicine initiatives.

+ Strengthening healthcare systems to ensure equitable distribution of resources, reduce healthcare costs, and improve the quality of care delivered to marginalized communities.

Overall, addressing racial and ethnic health disparities requires a multidimensional approach that recognizes the historical context, socio-economic factors, and systemic injustices that perpetuate these disparities. By implementing evidence-based interventions and policies aimed at reducing disparities, we can work towards achieving health equity and a more just society for all.

Structural Racism and Health Outcomes

Structural racism refers to the systemic and institutionalized ways in which racial inequities are perpetuated through policies, practices, and norms within society. In the context of health and wellness, structural racism encompasses the social, economic, and political factors that contribute to health disparities among different racial and ethnic groups. This section explores the impact of structural racism on health outcomes and discusses the ethical considerations associated with addressing this issue.

Understanding Structural Racism

Structural racism operates at multiple levels, including individual, interpersonal, and institutional levels. At the individual level, racial bias and discrimination can affect healthcare providers' perceptions, decisions, and behaviors towards patients of different racial and ethnic backgrounds. Interpersonal racism refers to the prejudice and discrimination experienced by individuals in their daily interactions with others, which can contribute to psychological stress and negatively impact health.

Institutional racism involves policies, practices, and norms within organizations and systems that perpetuate racial inequities. Examples include racially biased hiring practices, discriminatory lending practices that limit access to resources and opportunities, and segregation in housing and education. These systemic inequities can lead to disparities in access to quality healthcare, education, employment, and other social determinants of health.

Impact on Health Outcomes

Structural racism has profound effects on health outcomes, resulting in significant disparities in morbidity, mortality, and overall well-being among racial and ethnic groups. For example, studies have shown that Black Americans experience higher rates of chronic conditions such as hypertension, diabetes, and cardiovascular disease compared to their White counterparts. Additionally, infant mortality rates and maternal mortality rates are higher among Black women compared to White women.

These health disparities are the result of a complex interplay of factors influenced by structural racism. Limited access to quality healthcare services, including preventive care and early intervention, plays a significant role in perpetuating these disparities. Moreover, social determinants of health, such as education, employment, housing, and neighborhood characteristics, are all influenced by structural racism and contribute to health inequities.

Ethical Considerations

Addressing structural racism in healthcare and promoting health equity requires ethical considerations that prioritize justice, fairness, and autonomy. Here are key ethical considerations associated with addressing structural racism:

1. **Recognition and Acknowledgment** : Recognizing and acknowledging the existence of structural racism is crucial. Healthcare providers and policymakers must acknowledge the historical and current inequities faced by marginalized communities and commit to addressing them.

2. **Cultural Competence and Anti-racist Practices** : Healthcare professionals should strive to develop cultural competence, which involves understanding and respecting diverse cultural backgrounds and tailoring care to meet the unique needs of patients. Additionally, anti-racist practices within healthcare institutions involve actively working to dismantle systemic racism and promoting inclusivity.

3. **Health Equity in Policies and Programs** : Policymakers and public health officials should develop and implement policies and programs that promote health equity by addressing the underlying social determinants of health affected by structural racism. This may include increasing access to quality healthcare, improving educational opportunities, and reducing economic disparities.

4. **Community Engagement and Empowerment** : Engaging and empowering communities affected by structural racism is essential. Including community members in the decision-making process, implementing community-led initiatives, and collaborating with grassroots organizations can help ensure that interventions are effective, culturally appropriate, and address the specific needs of marginalized populations.

Case Study: Structural Racism and COVID-19 Disparities

The COVID-19 pandemic has highlighted the profound impact of structural racism on health outcomes. Communities of color, particularly Black, Indigenous, and Latinx populations, have been disproportionately affected by the pandemic. This disparity is attributed to a combination of factors, including higher rates of underlying health conditions, limited access to healthcare, and a higher prevalence of essential workers in these communities.

Structural racism plays a significant role in these disparities, as it contributes to the social and economic conditions that increase the risk of exposure and severe illness. Racism within healthcare systems, such as racial bias in treatment decisions and inadequate representation of racial minorities in clinical trials, further exacerbate these disparities.

Addressing these disparities requires a comprehensive approach that considers the underlying social determinants of health affected by structural racism. This includes expanding access to testing and healthcare, addressing racial bias within healthcare systems, providing economic support for affected communities, and ensuring equitable vaccine distribution.

Conclusion

Structural racism significantly impacts health outcomes and perpetuates health disparities among different racial and ethnic groups. Addressing this issue requires recognizing and acknowledging the existence of structural racism, promoting cultural competence and anti-racist practices within healthcare, prioritizing health equity in policies and programs, and engaging and empowering affected communities. By taking these ethical considerations into account, we can work towards a more just and equitable health and wellness future for all.

Implicit Bias in Healthcare

Implicit bias refers to the unconscious attitudes, beliefs, stereotypes, and prejudices that individuals hold towards certain social groups. These biases are activated involuntarily and without conscious awareness, and they can significantly influence our perceptions, decisions, and behavior towards others.

In the context of healthcare, implicit bias can have serious implications for patient outcomes, particularly for marginalized and minority populations. It can lead to unequal treatment, disparities in healthcare access and quality, and ultimately, worsened health outcomes.

Understanding Implicit Bias

Implicit bias is rooted in our socialization and cultural experiences. It operates at a subconscious level and can shape our judgments and actions, even when we consciously reject stereotypes or prejudices. These biases can be both positive and negative, but the negative biases are of particular concern in healthcare settings.

Research has shown that healthcare providers, like all individuals, can hold implicit biases based on race, ethnicity, gender, age, weight, and other social categories. For example, a study found that medical professionals exhibit implicit biases that associate Black individuals with lower pain tolerance, leading to under-treatment of pain in Black patients.

The Impact of Implicit Bias in Healthcare

Implicit bias can affect healthcare providers' decision-making in several ways:

+ Diagnosis and Treatment: Implicit biases can influence providers' diagnostic decisions, leading to misdiagnosis or delayed diagnosis for certain patient populations. For example, biases towards believing that women are more emotional or exaggerate symptoms may lead to the dismissal or undertreatment of their ailments.

+ Communication and Patient Trust: Implicit biases can affect provider-patient communication, leading to misunderstandings, decreased patient satisfaction, and reduced trust in the healthcare system. Patients who perceive bias may be less likely to fully disclose their symptoms or follow recommended treatment plans.

+ Treatment Recommendations: Implicit biases can impact treatment recommendations, resulting in differential access to appropriate care. For example, biases towards assuming lower socioeconomic status in certain racial or ethnic groups may influence providers to recommend less expensive treatment options without considering the patient's preferences or clinical indications.

+ Health Disparities: Cumulative effects of implicit bias contribute to health disparities, as marginalized populations often face unequal healthcare access, poorer quality of care, and higher rates of disease and mortality.

Addressing Implicit Bias

Addressing implicit bias is crucial for promoting equitable healthcare. Here are some strategies to mitigate the impact of implicit bias:

+ Awareness and Education: Healthcare providers should undergo training to raise awareness about implicit biases and their potential impact on patient care. This training can include recognizing unconscious biases, understanding their consequences, and learning strategies to overcome them.

+ Culturally Competent Care: Adopting culturally competent care practices is vital. Healthcare providers should strive to understand diverse cultural backgrounds, beliefs, and values to provide patient-centered care that respects individual differences.

+ Bias-Mitigating Interventions: Implementing interventions at the systemic level can help address implicit bias. Examples include implementing standardized protocols and guidelines for diagnosis and treatment, diverse representation in healthcare leadership, and promoting diversity and inclusion within the healthcare workforce.

+ Data Collection and Monitoring: Collecting and analyzing data on healthcare disparities and outcomes can help identify and address bias-related disparities. Monitoring and reporting these disparities allow for monitoring progress and holding healthcare systems accountable.

Case Study: Implicit Bias in Maternal Healthcare

Maternal healthcare provides an example of how implicit bias can impact healthcare outcomes. Studies have shown that Black women in the United States experience higher rates of maternal mortality and severe maternal morbidity compared to white women, even after controlling for socioeconomic factors.

Implicit bias can contribute to these disparities. For instance, biases towards perceiving Black women as less intelligent or less educated may influence healthcare providers to dismiss their concerns or provide suboptimal care. Additionally, stereotypes about pain tolerance and resilience may lead to inadequate pain management during childbirth.

To address implicit bias in maternal healthcare, interventions such as anti-bias training for healthcare providers, diverse representation in decision-making positions, and patient advocacy initiatives have been implemented. These efforts aim to promote equitable access to quality maternal healthcare and reduce disparities.

Conclusion

Implicit bias in healthcare is a significant ethical concern that can perpetuate healthcare disparities and inequities. Recognizing the presence and impact of implicit biases is the first step towards addressing them. By promoting awareness, education, and implementing systemic changes, healthcare systems can work towards providing equitable, just, and high-quality care for all individuals, regardless of their social identities. It is essential that healthcare professionals continuously reflect on their biases and strive to provide unbiased care to ensure optimal patient outcomes.

Culturally Competent Care

In the context of healthcare, culturally competent care is an essential aspect of providing equitable and effective health services to individuals from different cultural backgrounds. It involves understanding and respecting the diverse beliefs, values, customs, and practices of patients, and tailoring healthcare delivery to meet their specific cultural needs. Culturally competent care aims to eliminate disparities in health outcomes by addressing the social, cultural, and linguistic factors that may affect access to and quality of healthcare.

The Importance of Culturally Competent Care

Cultural competence recognizes that healthcare is not a one-size-fits-all approach. Different cultures have unique perspectives on health, illness, wellness, and healing, which may impact their attitudes towards healthcare and treatment options. By incorporating cultural knowledge and sensitivity into healthcare practice, providers can establish trust, enhance patient engagement, and improve health outcomes.

Culturally competent care has numerous benefits, such as:

1. Improved patient satisfaction: When healthcare providers understand and respect patients' cultural beliefs and practices, it can lead to greater patient satisfaction with their care experience.

2. Increased patient trust and engagement: Patients are more likely to trust and actively participate in their healthcare when they feel that their cultural background is respected by the healthcare providers.

3. Enhanced communication: Culturally competent care encourages effective communication between healthcare professionals and patients, leading to increased understanding, compliance with treatment plans, and better health outcomes.

4. Reduced health disparities: By addressing cultural barriers and tailoring care to the specific needs of different cultural groups, healthcare providers can help reduce health disparities that may exist among marginalized populations.

Principles of Culturally Competent Care

The principles of culturally competent care provide guidance for healthcare providers to effectively deliver healthcare services across diverse cultural contexts. These principles include:

1. Cultural knowledge: Healthcare providers should seek to understand the cultural beliefs, values, traditions, and social determinants of health that influence patient care decisions and health outcomes.

2. Sensitivity and awareness: It is crucial for healthcare professionals to be aware of their own cultural biases and prejudices, and be sensitive to the unique needs and experiences of individuals from different cultures.

3. Trust and rapport: Building trust and establishing a rapport with patients is essential for effective healthcare delivery. Healthcare providers should create a safe and non-judgmental environment where patients feel comfortable expressing their cultural beliefs and concerns.

4. Effective communication: Healthcare providers should strive for effective cross-cultural communication by using interpreters or bilingual staff, employing clear and simple language, and actively listening to patients' concerns.

5. Inclusion and cultural humility: Cultural humility involves recognizing the limitations of one's own cultural perspectives and adopting a stance of openness and willingness to learn from others. Healthcare providers should embrace diverse cultural practices, allowing patients to actively participate in decisions about their care.

Challenges and Strategies for Culturally Competent Care

While culturally competent care is crucial, it also brings challenges that healthcare providers may encounter. Some of these challenges include:

1. Language barriers: Communication can be hindered when patients and healthcare providers do not share a common language. To overcome this challenge, healthcare organizations can provide interpreter services or employ bilingual staff members.

2. Stereotyping and bias: Stereotyping and bias can influence the assumptions and judgments healthcare providers make about patients from different cultures. To mitigate this challenge, healthcare professionals should engage in self-reflection and

ongoing cultural competency training to minimize the impact of stereotypes and biases on patient care.

3. Limited cultural knowledge: Healthcare providers may have limited knowledge about certain cultural practices, beliefs, or health traditions. Continuous education and cultural competency training can help fill this knowledge gap.

4. Time constraints: Healthcare providers often face time constraints, making it challenging to provide culturally competent care. Strategies such as incorporating cultural assessment tools into electronic health records and utilizing support staff trained in cultural competency can help ensure time-efficient delivery of culturally competent care.

In order to promote culturally competent care, healthcare organizations and providers can implement the following strategies:

1. Cultural competency training: Healthcare professionals should receive ongoing training and education on cultural competence to enhance their understanding of diverse cultural practices and beliefs.

2. Diverse healthcare workforce: Hiring a diverse healthcare workforce can help bridge the cultural gap between providers and patients, and improve overall cultural competence within organizations.

3. Community collaboration: Building partnerships with community organizations and cultural leaders can help healthcare providers better understand the needs and healthcare practices of specific cultural groups.

4. Culturally sensitive healthcare policies: Healthcare organizations should develop policies that promote cultural competence and address health disparities within different cultural communities.

Case Study: Culturally Competent Mental Health Care

Let's consider a case study to illustrate the importance of culturally competent care in mental health:

Maria, a Hispanic woman, seeks mental health treatment for symptoms of depression and anxiety. The healthcare provider, Anna, who lacks cultural knowledge and sensitivity, assumes that Maria's symptoms are related to personal weakness rather than considering the potential cultural factors influencing her mental health. Anna prescribes medication without discussing alternative treatment options or considering Maria's cultural background.

In this scenario, Anna's lack of cultural competence may lead to poor patient outcomes and dissatisfaction with the care provided. To provide culturally competent care, Anna should:

1. Seek cultural knowledge: Anna should familiarize herself with the cultural beliefs, values, and healing practices of the Hispanic community, particularly related to mental health. This knowledge will help her understand the unique cultural factors that may contribute to Maria's symptoms.

2. Build trust and rapport: Anna should create a safe and welcoming environment for Maria to disclose her concerns and cultural background. Developing a trusting relationship will facilitate open communication and ensure that Maria's unique cultural needs are addressed.

3. Collaborate on treatment plans: Anna should engage Maria in shared decision-making, considering her preferences and cultural beliefs when formulating a treatment plan. This collaborative approach will increase Maria's engagement and adherence to the recommended treatment.

4. Utilize cultural resources: Anna can collaborate with cultural experts or community organizations to better understand the cultural nuances impacting mental health within the Hispanic community. These resources can provide Anna with the necessary tools to deliver culturally sensitive care.

By adopting culturally competent care practices, Anna can improve the quality of care provided to Maria, leading to better mental health outcomes and patient satisfaction.

Conclusion

Culturally competent care is essential for achieving equitable and effective healthcare delivery. Healthcare professionals should strive to understand and respect the cultural beliefs, values, and practices of their patients. By incorporating cultural knowledge into healthcare practices, building trust, and adopting culturally sensitive strategies, healthcare providers can promote better health outcomes and reduce healthcare disparities among diverse cultural populations. Embracing cultural competence is a crucial step towards creating a more equitable, just, and compassionate health and wellness future.

Gender Inequalities in Health

Gender Identity and Health

Gender identity plays a significant role in shaping an individual's overall health and well-being. It refers to a person's deeply-held sense of their own gender, which may not necessarily align with the sex they were assigned at birth. Understanding the impact of gender identity on health is essential for promoting equitable and inclusive healthcare systems.

The Importance of Gender-affirming Care

Gender-affirming care is a critical component of providing comprehensive healthcare for individuals with diverse gender identities. This approach, also known as transgender-affirming care, involves affirming and respecting an individual's self-identified gender and addressing their unique health needs.

Gender-affirming care encompasses a range of medical, psychological, and social interventions. Some of the key components include:

+ Transition-related healthcare: This includes hormone therapy, surgical interventions (such as gender-affirming surgeries), voice therapy, and other procedures that help individuals align their physical attributes with their gender identity.

+ Mental healthcare: Many individuals with diverse gender identities face higher rates of mental health challenges, including depression, anxiety, and gender dysphoria. Access to mental healthcare providers who are knowledgeable and sensitive to these concerns is crucial.

+ Social support: Creating a supportive environment is vital for the health and well-being of individuals with diverse gender identities. This can involve support groups, community organizations, and policies that protect against discrimination and promote inclusivity.

It is essential for healthcare providers to be aware of the unique needs and challenges faced by individuals with diverse gender identities. Mental health professionals play a crucial role in providing support and therapeutic interventions to address gender dysphoria and related mental health concerns.

Health Disparities

Despite progress in recognizing and understanding gender diversity, individuals with diverse gender identities continue to face significant health disparities. These disparities are often compounded by factors such as race, socioeconomic status, and disability.

Some of the key health disparities faced by individuals with diverse gender identities include:

+ **Higher rates of mental health challenges:** Studies have consistently shown that transgender and non-binary individuals experience higher rates of depression, anxiety, self-harm, and suicidal ideation compared to the general population. These disparities are often linked to experiences of discrimination, stigma, and social isolation.

+ **Limited access to gender-affirming care:** Many healthcare systems have barriers that restrict access to gender-affirming care, including hormone therapy and gender-affirming surgeries. Limited access to these interventions can negatively impact the physical and mental well-being of individuals with diverse gender identities.

+ **Increased rates of violence:** Transgender and gender non-binary individuals are more likely to experience violence, including hate crimes and intimate partner violence. This violence can have severe physical and psychological consequences and further exacerbate health disparities.

+ **Healthcare discrimination:** Discrimination in healthcare settings is a significant barrier to accessing quality and comprehensive care for individuals with diverse gender identities. Misgendering, refusal of care, and lack of cultural competence among healthcare providers can create significant challenges and barriers to care.

Addressing these health disparities requires a multi-faceted approach that includes policy changes, education and training for healthcare providers, community support, and advocacy for the rights of individuals with diverse gender identities.

Healthcare Provider Responsibilities

Healthcare providers have a crucial role to play in promoting the health and well-being of individuals with diverse gender identities. Some key responsibilities include:

+ **Cultural competence:** Healthcare providers should receive training on the unique health needs and experiences of individuals with diverse gender identities. This includes understanding appropriate language, practicing inclusive communication, and respecting the individual's self-identified gender.

+ **Creating a safe and inclusive environment:** Healthcare settings should be safe spaces where individuals feel comfortable discussing their gender identity and expressing their healthcare needs. This involves implementing policies that protect against discrimination, ensuring privacy and confidentiality, and addressing any biases or prejudices.

+ **Providing informed and affirming care:** Healthcare providers should be knowledgeable about the best practices for gender-affirming care, including hormone therapy, surgical options, mental health support, and preventive care. They should be prepared to offer evidence-based care that respects and supports the individual's gender identity.

+ **Advocacy and support:** Healthcare providers can play a vital role in advocating for the rights and well-being of individuals with diverse gender identities. This can involve supporting policy changes, participating in professional organizations focused on gender diversity, and providing resources and referrals to community organizations.

By fulfilling these responsibilities, healthcare providers can contribute to reducing health disparities and promoting the overall health and well-being of individuals with diverse gender identities.

Case Example: Addressing Health Disparities

Let's consider the case of Alex, a transgender man who has been avoiding seeking healthcare due to fear of discrimination and lack of understanding. Alex has been experiencing symptoms of depression and anxiety but has not received any professional support.

In this case, a healthcare provider who is trained in transgender healthcare can make a significant difference. By creating a safe and inclusive environment, the provider can help Alex feel comfortable discussing his mental health concerns. The provider can also ensure that appropriate language and communication practices are used, validating Alex's gender identity and avoiding any misgendering.

The healthcare provider can then provide informed and affirming care by conducting a thorough assessment of Alex's mental health symptoms, considering both the gender-related factors and other potential causes. With a comprehensive understanding of Alex's needs, the provider can offer appropriate interventions, such as therapy or medication, tailored to Alex's experiences as a transgender man.

Additionally, the healthcare provider can advocate for Alex by ensuring that he has access to inclusive mental health resources, support groups, and community organizations. They can also educate other healthcare professionals and contribute to policy changes that protect the rights and well-being of individuals with diverse gender identities.

Through these steps, the healthcare provider can address the health disparities faced by individuals like Alex, promoting improved mental health outcomes and overall well-being.

Key Takeaways

+ Gender identity significantly impacts an individual's health and well-being. Gender-affirming care is crucial for providing comprehensive healthcare to individuals with diverse gender identities.

+ Individuals with diverse gender identities face significant health disparities, including higher rates of mental health challenges, limited access to gender-affirming care, increased rates of violence, and healthcare discrimination.

+ Healthcare providers have a responsibility to provide culturally competent and affirming care to individuals with diverse gender identities. This involves creating a safe and inclusive environment, providing informed care, and advocating for their rights and well-being.

+ Addressing health disparities requires a multi-faceted approach that includes policy changes, education and training, community support, and advocacy.

+ Case example: By providing culturally competent and affirming care, healthcare providers can address the unique needs of individuals with diverse gender identities and contribute to improving their health outcomes.

Exercises

1. Research and discuss the healthcare disparities faced by transgender and gender non-binary individuals in your country or region. What are the underlying factors contributing to these disparities? What steps can be taken to address them?

2. Identify a healthcare setting or organization in your community that is known for providing gender-affirming care. Interview a healthcare provider from that setting to understand their approach to providing inclusive care. Share your findings and reflect on the impact of such care on individuals with diverse gender identities.

Resources

+ World Professional Association for Transgender Health (WPATH) - https://www.wpath.org/

+ The Trevor Project - https://www.thetrevorproject.org/

+ National LGBTQIA+ Health Education Center - https://www.lgbtqiahealtheducation.org/

+ Gender Diversity - https://www.genderdiversity.org/

+ Transgender Legal Defense and Education Fund - https://transgenderlegal.org/

Women's Reproductive Health Rights

Women's reproductive health rights encompass a broad range of issues related to women's autonomy, agency, and access to reproductive healthcare. In this section, we will explore the fundamental principles and ethical considerations surrounding women's reproductive health rights, including abortion, contraception, and maternal healthcare.

The Importance of Women's Reproductive Health Rights

Women's reproductive health rights are essential for ensuring gender equality, bodily autonomy, and the overall well-being of women. These rights recognize that women have the right to control their reproductive choices, including the decision to have children, the spacing of pregnancies, and the right to access safe and legal abortion services. Reproductive health rights are also crucial for addressing issues such as maternal mortality and morbidity, gender-based violence, and ensuring women's access to essential healthcare services.

Abortion and Access to Reproductive Healthcare

Abortion is a highly contentious and complex issue from both ethical and legal perspectives. The debate revolves around the conflicting values of women's autonomy and the rights of the fetus. Ethical frameworks used to examine this issue include reproductive rights, bodily autonomy, the right to privacy, and the principles of justice and fairness.

From a reproductive rights perspective, women have the right to make decisions about their own bodies, including the decision to terminate a pregnancy. This perspective emphasizes the importance of ensuring access to safe and legal abortion services, as restrictions on abortion can lead to unsafe practices and harm women's health.

The principle of bodily autonomy also plays a central role in the abortion debate. It asserts that women have the right to control what happens to their own bodies, including decisions about pregnancy and childbirth. Restrictions on abortion infringe upon this principle and can violate women's rights.

Moreover, the right to privacy is often invoked in discussions about abortion. This right protects a woman's personal decision-making regarding her reproductive choices, without unwarranted interference from the state or other individuals.

Addressing the principles of justice and fairness is also crucial in the context of abortion. These principles emphasize the importance of ensuring equitable access to reproductive healthcare services for all women, regardless of socio-economic status or geographical location.

Contraception and Family Planning

Access to contraception is a vital component of women's reproductive health rights. Contraception enables women to prevent unintended pregnancies, improve maternal and child health outcomes, and exercise greater control over their reproductive choices.

Ethical considerations surrounding contraception primarily revolve around the principles of autonomy, privacy, and justice. Autonomy allows women to make informed decisions about their contraceptive methods based on their personal values, beliefs, and health needs. Privacy ensures that women have the right to access and use contraception without fear of stigmatization or discrimination. Justice demands that contraception be available and affordable to all women, regardless of their socio-economic status.

However, barriers to contraception persist, including limited access, cost, cultural and religious beliefs, and inadequate education. Addressing these barriers and promoting comprehensive sexuality education are vital for upholding women's reproductive health rights.

Maternal Healthcare and Maternity Leave

Maternal healthcare plays a crucial role in protecting women's reproductive health rights. Adequate prenatal care, skilled birth attendants, and access to postpartum care are essential to ensure safe and healthy pregnancies and childbirth.

Ethically, maternal healthcare is grounded in the principles of beneficence and non-maleficence. Women have the right to receive care that promotes their well-being and minimizes harm during childbirth. Health systems must prioritize the provision of evidence-based and culturally sensitive maternal healthcare to reduce maternal mortality and morbidity rates.

An important aspect of supporting women's reproductive health rights is providing adequate maternity leave. Maternity leave allows women to recover from childbirth, bond with their newborns, and continue breastfeeding. It also promotes the principle of justice by recognizing the unique needs of women during the postpartum period.

However, women's access to quality maternal healthcare and maternity leave can vary significantly depending on socio-economic disparities, geographical location, and cultural norms. Addressing these inequalities is crucial for promoting women's reproductive health rights and ensuring optimal maternal and child health outcomes.

Addressing Challenges and Advancing Women's Reproductive Health Rights

Advocacy, education, policy changes, and a human rights-based approach are essential for advancing women's reproductive health rights. It is crucial to challenge myths, misinformation, and stigma surrounding reproductive health issues and to promote accurate and comprehensive sexual and reproductive health education.

Efforts should focus on improving access to reproductive healthcare services, including safe and legal abortion, contraception, and maternal healthcare. This requires removing financial, geographical, and legal barriers and ensuring equitable access for all women, including marginalized and vulnerable populations.

Furthermore, involving women in decision-making processes, including policy development and implementation, is vital for ensuring that their voices are heard and their rights are protected.

In conclusion, women's reproductive health rights are crucial for promoting gender equality, autonomy, and overall well-being. Upholding these rights requires addressing ethical considerations surrounding abortion, contraception, and maternal healthcare and working towards creating a society that respects and supports women's reproductive choices and agency.

Key Takeaways

- Women's reproductive health rights encompass a range of issues related to women's autonomy, agency, and access to reproductive healthcare.

- Ethical considerations surrounding women's reproductive health rights include principles of autonomy, bodily autonomy, privacy, justice, and fairness.

- Abortion and access to safe and legal abortion services are central issues in women's reproductive health rights, with competing perspectives on women's autonomy and the rights of the fetus.

- Access to contraception is essential for enabling women to make informed decisions about their reproductive choices and exercising their autonomy.

- Maternal healthcare and adequate maternity leave are crucial components of women's reproductive health rights, ensuring safe and healthy pregnancies and childbirth.

- Addressing challenges and advancing women's reproductive health rights require advocacy, education, policy changes, and a human rights-based approach.

Discussion Questions

1. How do cultural and religious beliefs influence women's access to reproductive healthcare services?

2. What are some effective strategies for promoting comprehensive sexuality education and reducing stigma surrounding reproductive health issues?

3. How can policymakers ensure equitable access to safe and legal abortion services?

4. What steps can be taken to address socio-economic disparities in access to contraception and maternal healthcare?

5. How can healthcare systems support breastfeeding and promote maternal mental health during the postpartum period?

Gender-based Violence and Health

Gender-based violence is a pervasive issue that affects individuals worldwide, with significant implications for their physical, mental, and social well-being. In this section, we will explore the ethical considerations surrounding gender-based violence and its impact on health. We will discuss the root causes of gender-based violence, the consequences for survivors, and the strategies for prevention and support.

Understanding Gender-based Violence

Gender-based violence refers to any act that is perpetrated against an individual based on their gender, resulting in physical, sexual, psychological, or economic harm. It is rooted in gender inequality and reinforces power imbalances between men and women, perpetuating harmful norms and stereotypes.

There are various forms of gender-based violence, including intimate partner violence, sexual assault, rape, female genital mutilation, forced marriage, and harassment. These acts can occur in public or private spaces, and they are fueled by factors such as patriarchal norms, societal expectations, and harmful cultural practices.

The Impact on Health

Gender-based violence has profound consequences for the health and well-being of survivors. The physical injuries resulting from violence can range from bruises and broken bones to severe internal injuries and even death. Survivors may also experience long-term health issues such as chronic pain, sexually transmitted infections, gynecological problems, and pregnancy complications.

The psychological impact of gender-based violence is significant. Survivors often experience post-traumatic stress disorder, anxiety, depression, and a diminished sense of self-worth. These mental health issues can have long-lasting effects on individuals' overall well-being and quality of life.

Additionally, gender-based violence can contribute to social isolation, economic instability, and reduced access to healthcare. Survivors may face barriers in seeking help and support due to stigma, fear of retaliation, or lack of resources.

Ethical Considerations

Addressing gender-based violence requires a multifaceted approach grounded in ethical principles. Some key ethical considerations include:

Respect for Autonomy: Respecting the autonomy and agency of survivors is crucial. It is essential to provide survivors with information, support, and resources, empowering them to make informed decisions about their safety and well-being.

Non-maleficence: Preventing further harm is paramount. Efforts should focus on interrupting the cycle of violence, protecting survivors, and holding perpetrators accountable.

Justice and Equity: Promoting justice and equity is fundamental to addressing gender-based violence. This includes combating social norms that perpetuate violence, advocating for legal protections, and promoting gender equality in all spheres of life.

Confidentiality and Privacy: Respect for confidentiality and privacy is vital in supporting survivors. Confidentiality should be prioritized when providing healthcare, counseling, and support services to ensure that survivors feel safe and empowered to seek help.

Prevention and Support Strategies

Preventing and addressing gender-based violence requires a comprehensive approach involving various stakeholders, including individuals, communities, governments, and organizations. Here are some strategies for prevention and support:

Education and Awareness: Promoting education and awareness is key to challenging harmful gender norms and stereotypes. By engaging in conversations about consent, healthy relationships, and gender equality, we can foster a culture of respect and prevent gender-based violence.

Legislation and Policy: Implementing and enforcing legislation that criminalizes gender-based violence is crucial. Laws should protect survivors, hold perpetrators accountable, and ensure access to justice and support services.

Support Services: Establishing accessible and comprehensive support services is essential for survivors. This includes safe shelters, helplines, counseling, legal aid, and healthcare services.

Community Mobilization: Engaging communities in the prevention of gender-based violence is crucial. Community-based initiatives can challenge harmful norms, provide social support, and create safe spaces for survivors.

Capacity Building: Building the capacity of healthcare providers, law enforcement, and other professionals is essential. Training should focus on recognizing and responding to gender-based violence sensitively and effectively.

Case Study: The Me Too Movement

The Me Too movement, which gained global momentum in 2017, serves as a powerful example of how survivors can amplify their voices and bring attention to the prevalence of gender-based violence. It began as a social media campaign, encouraging individuals to share their experiences with sexual harassment and assault.

The movement sparked a public dialogue about consent, power dynamics, and accountability. It prompted widespread awareness and generated discussions about the need for systemic change to prevent and address gender-based violence. The Me Too movement demonstrates the power of collective action in challenging societal norms and advocating for justice and healing.

Conclusion

Gender-based violence is a complex issue that requires a comprehensive and ethical response. By addressing the root causes, promoting gender equality, and supporting survivors, we can work towards a world free from gender-based violence. It is crucial to actively engage in prevention, support survivors, and advocate for policies that promote justice and equality. Together, we can create a society that values the health, well-being, and dignity of all individuals, irrespective of their gender.

LGBTQ+ Health Disparities

In this section, we will explore the ethical issues surrounding the health disparities faced by the LGBTQ+ community. LGBTQ+ individuals often experience significant barriers to accessing healthcare and suffer from higher rates of physical and mental health problems compared to their heterosexual and cisgender counterparts. Understanding the root causes of these disparities, addressing them with ethical principles and strategies, and promoting inclusive and equitable healthcare are crucial steps towards achieving the goal of health and wellness for all individuals, regardless of sexual orientation or gender identity.

Background

The LGBTQ+ community encompasses individuals who identify as lesbian, gay, bisexual, transgender, queer, and other diverse sexual orientations and gender identities. Historically, LGBTQ+ individuals have faced discrimination, marginalization, and stigmatization, which have contributed to their health disparities. These disparities can manifest in various ways, including higher rates of mental health disorders, substance abuse, HIV/AIDS, and other sexually transmitted infections (STIs), as well as limited access to healthcare services.

Principles of Ethical Considerations

To address LGBTQ+ health disparities, it is necessary to consider several ethical principles:

1. **Autonomy and Informed Consent:** LGBTQ+ individuals should have the right to make informed decisions about their healthcare, free from coercion or discrimination.

2. **Justice and Equity:** Healthcare should be provided in a fair and equitable manner, ensuring equal access and treatment for LGBTQ+ individuals.

3. **Non-Maleficence:** Healthcare professionals should not harm LGBTQ+ individuals through actions or omissions that perpetuate discrimination or deny them appropriate care.

4. **Beneficence:** Efforts should be made to promote the well-being and improve the health outcomes of LGBTQ+ individuals.

5. **Cultural Competence:** Healthcare providers should possess the knowledge, attitudes, and skills necessary to provide culturally sensitive care to LGBTQ+ individuals.

Adhering to these ethical principles can guide healthcare professionals and policymakers towards creating inclusive and equitable healthcare systems for LGBTQ+ individuals.

Health Disparities faced by LGBTQ+ Individuals

LGBTQ+ individuals experience higher rates of several health disparities compared to the general population. These disparities can be attributed to a range of factors, including social stigma, discrimination, inadequate healthcare infrastructure, and limited provider knowledge and training. Let's explore some key health disparities faced by LGBTQ+ individuals:

1. **Mental Health Disparities:** LGBTQ+ individuals are more likely to experience mental health disorders such as depression, anxiety, and suicidal ideation. The stigma and discrimination they face contribute to increased psychological distress.

2. **Sexual Health Disparities:** LGBTQ+ individuals have higher rates of STIs, including HIV/AIDS. Factors such as lack of comprehensive sex education, limited access to healthcare, and stigma surrounding sexual orientation and gender identity contribute to these disparities.

3. **Substance Abuse:** LGBTQ+ individuals have higher rates of substance abuse, including alcohol, tobacco, and drug use. This can be due to the coping mechanisms individuals employ to deal with minority stress, discrimination, and the challenges associated with their sexual orientation or gender identity.

4. **Cancer Disparities:** Certain types of cancers, such as breast and cervical cancer, disproportionately affect LGBTQ+ individuals. Barriers to timely screenings, lack of healthcare access, and provider discrimination contribute to these disparities.

5. **Limited Access to Gender-Affirming Care:** Transgender and gender non-conforming individuals often face challenges in accessing gender-affirming healthcare services, including hormone therapy and gender-affirming surgeries. Provider bias, lack of knowledgeable providers, and insurance coverage limitations can impede access to essential care.

Strategies and Solutions

To address LGBTQ+ health disparities, it is important to implement strategies that promote inclusive and equitable healthcare. Here are some key strategies:

1. **Education and Training:** Healthcare providers should receive comprehensive training on LGBTQ+ health issues, cultural competence, and inclusive practices. This will help improve understanding, reduce bias, and enhance the quality of care provided to LGBTQ+ individuals.

2. **Policy Changes:** Implementation of policies that protect against discrimination based on sexual orientation and gender identity is crucial. These policies can foster a safe and inclusive environment for LGBTQ+ individuals, both within healthcare settings and in society at large.

3. **Community Engagement:** Collaboration with LGBTQ+ organizations and community leaders is essential to identify and address the specific healthcare needs of the LGBTQ+ community. Establishing partnerships can help ensure that services are accessible, culturally competent, and tailored to the needs of LGBTQ+ individuals.

4. **Research and Data Collection:** More research is needed to understand the unique health needs and experiences of LGBTQ+ individuals. Collecting sex, gender, and sexual orientation data in healthcare settings can help identify disparities and guide the development of targeted interventions.

5. **Creating Safe Spaces:** Healthcare settings should strive to create safe and affirming spaces for LGBTQ+ individuals. This includes ensuring confidentiality, respecting chosen names and pronouns, displaying LGBTQ+-inclusive signage and literature, and training staff to provide respectful and non-discriminatory care.

Real-world Example

One real-world example of addressing LGBTQ+ health disparities is the implementation of the "Affirming Care Model" at a community health center. The model incorporates comprehensive staff training on LGBTQ+ health issues, cultural competence, and inclusive language. It includes creating safe spaces, instituting LGBTQ+-inclusive policies, and establishing partnerships with local LGBTQ+ organizations to improve healthcare access and reduce disparities. The model has been successful in increasing patient satisfaction, reducing healthcare disparities, and improving health outcomes for LGBTQ+ individuals in the community.

Conclusion

Addressing LGBTQ+ health disparities requires a multifaceted approach that integrates ethical principles, policy changes, education, community engagement, and research. By promoting inclusive and equitable healthcare, we can work towards eliminating disparities and ensuring that all individuals, regardless of sexual orientation or gender identity, have the opportunity to achieve optimal health and wellness. It is essential that healthcare providers, policymakers, and society as a whole recognize and act upon the unique healthcare needs and challenges faced by LGBTQ+ individuals, striving towards a future where everyone can access the care they need in a safe and affirming manner.

Environmental Determinants of Health and Wellness

Climate Change and Health

Climate change is a pressing global issue that has profound implications for human health and well-being. The Earth's climate is changing as a result of human activities, primarily the burning of fossil fuels and deforestation, leading to an increase in greenhouse gas emissions and subsequent global warming. This rise in global temperatures has far-reaching consequences for various aspects of human health, including physical, mental, and social well-being.

Understanding the Impact of Climate Change on Health

Climate change affects health through several direct and indirect pathways. The direct impact includes heat-related illnesses, exacerbated respiratory and cardiovascular conditions, and increased mortality rates during extreme weather events such as heatwaves, floods, and hurricanes. Indirectly, climate change influences health through altered ecological systems, changes in infectious disease patterns, food and water insecurity, displacement and migration, and social and economic disruption.

Climate Change and Infectious Diseases

Climate change alters the distribution and frequency of infectious diseases, posing significant health risks. For example, as temperatures rise, disease vectors such as mosquitoes are able to expand their geographic range, increasing the transmission of diseases like malaria, dengue fever, and Zika virus. Changes in precipitation patterns can also contribute to the spread of waterborne diseases like cholera and other diarrheal illnesses. Moreover, extreme weather events can lead to the displacement of populations, overcrowding in temporary shelters, and inadequate sanitation, creating favorable conditions for disease outbreaks.

Climate Change and Mental Health

Climate change also has profound implications for mental health and well-being. As individuals and communities experience the impacts of extreme weather events, loss of livelihoods, and displacement, there is an increased risk of mental health disorders such as anxiety, depression, post-traumatic stress disorder (PTSD), and substance abuse. The psychological distress associated with climate change can be exacerbated by the uncertainty and fear of future climate-related events. Vulnerable populations, including children, the elderly, and those with pre-existing mental health conditions, are particularly at risk.

Mitigating and Adapting to Climate Change for Health Promotion

To address the health impacts of climate change, a multi-faceted approach is needed that focuses on both mitigating climate change and adapting to its effects. Mitigation strategies aim to reduce greenhouse gas emissions and the drivers of climate change, such as transitioning to renewable energy sources, promoting energy efficiency, and implementing sustainable transportation and land-use practices. By reducing the magnitude and rate of climate change, mitigation measures can indirectly safeguard human health.

Adaptation strategies focus on building resilience and preparedness to minimize the negative health consequences of climate change. This includes developing early warning systems for extreme weather events, strengthening healthcare infrastructure and emergency response systems, implementing heatwave management plans, improving water and sanitation infrastructure, and enhancing vector control measures. Additionally, community-based interventions, education, and awareness programs can help individuals and communities adapt to the changing climate and mitigate the health risks associated with climate change.

Equity and Climate Justice in Health Response

It is crucial to address the health impacts of climate change through a lens of equity and justice. Vulnerable populations, including low-income communities, marginalized groups, and developing countries, are disproportionately affected by climate change due to factors such as limited access to resources, infrastructure, and healthcare. Therefore, health responses to climate change must prioritize equitable access to healthcare, essential services, and resources to ensure that the most vulnerable communities are not left behind.

In conclusion, climate change poses significant challenges to global health and necessitates comprehensive strategies to mitigate its impact and adapt to its effects.

By addressing climate change through a health lens and promoting equity and justice in health responses, we can strive towards a more sustainable and resilient future for the health and well-being of all individuals and communities.

Air and Water Pollution

Air and water pollution are critical environmental issues that have far-reaching impacts on human health and well-being. In this section, we will explore the causes and effects of air and water pollution, as well as the ethical considerations and potential solutions to mitigate these problems.

Causes of Air Pollution

Air pollution is primarily caused by the release of harmful substances into the Earth's atmosphere. Some of the major sources of air pollution include:

+ **Industrial Emissions:** Industrial activities, such as manufacturing and power generation, release large amounts of pollutants into the air. These can include greenhouse gases, particulate matter, volatile organic compounds (VOCs), and various toxic chemicals.

+ **Transportation:** The combustion of fossil fuels in vehicles is a significant contributor to air pollution. The exhaust emissions from cars, trucks, and airplanes contain pollutants such as nitrogen oxides (NOx), carbon monoxide (CO), and particulate matter.

+ **Burning of Fossil Fuels:** The burning of coal, oil, and natural gas for energy production and residential heating is a major source of air pollution. This releases pollutants including sulfur dioxide (SO_2), nitrogen oxides (NOx), and carbon dioxide (CO_2) into the atmosphere.

+ **Agricultural Practices:** Agricultural activities, such as livestock farming and the use of synthetic fertilizers and pesticides, contribute to air pollution. Livestock emit methane (a potent greenhouse gas), while the use of fertilizers and pesticides releases ammonia and other chemicals into the air.

+ **Construction and Demolition:** Construction sites and demolition activities can generate dust, construction waste, and emissions from heavy machinery, leading to localized air pollution.

Effects of Air Pollution

Air pollution has profound effects on both human health and the environment. The impacts of air pollution include:

+ **Respiratory and Cardiovascular Diseases:** Inhalation of pollutants in the air can cause or worsen respiratory conditions such as asthma, bronchitis, and chronic obstructive pulmonary disease (COPD). Air pollution is also linked to an increased risk of cardiovascular diseases, including heart attacks and strokes.

+ **Reduced Lung Function:** Prolonged exposure to air pollution can lead to a decline in lung function, especially in children and the elderly.

+ **Climate Change:** Certain air pollutants, such as carbon dioxide and methane, contribute to the greenhouse effect and global warming, causing climate change and its subsequent environmental impacts.

+ **Environmental Degradation:** Air pollution can harm ecosystems and natural resources by acidifying lakes and forests, damaging crops, and depleting the ozone layer.

Ethical Considerations

Addressing air pollution raises several ethical considerations:

+ **Environmental Justice:** Air pollution often disproportionately affects marginalized communities, exacerbating existing social inequalities. It is crucial to address this injustice and ensure that everyone has equal access to clean air and a healthy environment.

+ **Responsibility and Accountability:** Industrial and commercial entities that contribute significantly to air pollution should be held accountable for their actions. Ethical responsibility requires businesses to mitigate their emissions, adopt cleaner technologies, and support sustainable practices.

+ **Interconnectedness:** Air pollution is a global issue that transcends national boundaries. Collaborative efforts and ethical responsibility are required to address this problem collectively, considering the interconnectedness of the global community.

Mitigation Strategies

To combat air pollution, a combination of regulatory measures, technological advancements, and individual actions is necessary. Some potential mitigation strategies include:

+ **Transition to Renewable Energy:** Accelerating the shift from fossil fuels to renewable energy sources such as solar, wind, and hydropower can significantly reduce air pollution associated with energy production.

+ **Improved Industrial Practices:** Implementing stricter emissions standards, promoting cleaner production technologies, and encouraging pollution control measures can reduce industrial air pollution.

+ **Promotion of Sustainable Transportation:** Encouraging the use of electric vehicles, public transportation, and active modes of transport like walking and cycling can decrease air pollution from transportation.

+ **Regulation and Monitoring:** Strengthening regulations and monitoring systems to ensure compliance with air quality standards is crucial in mitigating pollution. Regular monitoring can also provide valuable data for targeted interventions.

+ **Awareness and Education:** Raising awareness about the health and environmental impacts of air pollution and promoting individual actions, such as reducing personal emissions and supporting sustainable practices, can contribute to mitigating the problem.

Case Study: Air Pollution in Delhi, India

The city of Delhi, India, has faced severe air pollution challenges in recent years. Factors such as industrial emissions, vehicular pollution, construction activities, and agricultural practices contribute to the city's poor air quality. The high concentration of particulate matter and harmful gases has led to significant health issues among the population, including respiratory diseases and increased mortality rates.

To address this problem, the government of Delhi and various stakeholders have implemented several measures. These include the introduction of cleaner fuel standards, promoting the use of compressed natural gas (CNG) for transportation, implementing stricter industrial emission norms, and adopting odd-even traffic schemes to reduce vehicular pollution.

Despite these efforts, air pollution in Delhi remains a significant challenge, necessitating ongoing ethical considerations and innovative solutions. The case of Delhi highlights the complexity of addressing air pollution and the need for sustained commitment from all stakeholders.

Conclusion

Air pollution poses a significant threat to human health, environmental well-being, and social justice. By understanding the causes, effects, and ethical considerations associated with air pollution, we can work towards implementing effective mitigation strategies. Through collective responsibility, innovation, and informed decision-making, we can create a future with cleaner air and a healthier planet for present and future generations.

Food Systems and Health

Food is a fundamental aspect of our lives, providing the nourishment and energy needed for our bodies to function properly. However, the food we consume is not simply a matter of personal choice; it is deeply intertwined with larger food systems that have profound impacts on our health and well-being. In this section, we will explore the ethical issues surrounding food systems and their implications for health.

Understanding Food Systems

Food systems encompass all the processes, activities, and infrastructure involved in the production, distribution, and consumption of food. They involve complex interactions between various actors such as farmers, food processors, distributors, retailers, and consumers. Understanding the dynamics of food systems is essential for addressing the ethical challenges they present.

Ethical Issues in Food Systems

1. **Food Security and Access:** One of the primary ethical concerns in food systems is ensuring food security and access for all individuals. Inequities in access to nutritious and affordable food can lead to malnutrition and other health problems, particularly in marginalized communities. Therefore, policies and interventions should aim to address these disparities and ensure the availability of healthy food options for everyone.

2. **Food Safety and Quality:** Another critical ethical consideration is the safety and quality of the food we consume. Contamination, improper handling,

and inadequate labeling can pose health risks, ranging from foodborne illnesses to allergenic reactions.   Ensuring strict food safety regulations and promoting transparency in food labeling are essential for protecting public health and empowering consumers to make informed choices.

3. **Sustainability and Environmental Impacts:** Food systems exert significant pressure on the environment, including deforestation, water pollution, biodiversity loss, and greenhouse gas emissions. It is crucial to adopt sustainable practices in agriculture, minimize food waste, and promote regenerative farming techniques to mitigate these environmental impacts. Ethical food systems should aim to meet the needs of the present without compromising the ability of future generations to meet their own needs.

4.   **Worker Rights and Welfare:**   The agricultural and food processing industries often rely on the labor of low-wage workers who face poor working conditions, exploitation, and limited access to healthcare. Ethical food systems should prioritize fair labor practices, ensuring workers' rights are protected, and their well-being is prioritized.

5. **Animal Welfare:** The treatment of animals in the food production system raises ethical concerns. Many animals are subjected to confinement, overcrowding, and inhumane practices in factory farming. Promoting animal welfare in food systems involves acknowledging animals' intrinsic value, reducing their suffering, and embracing more humane agricultural practices.

Addressing Ethical Issues

1. **Policy Interventions:** Governments play a vital role in addressing ethical issues in food systems through legislation and regulations. This can include measures to ensure food safety, promote sustainable farming practices, and protect workers' rights. Policy interventions can also target improving food access in underserved communities and support local food systems.

2.   **Consumer Awareness and Education:** Empowering consumers with knowledge about ethical food choices can drive demand for sustainable, ethically produced food.   Promoting education programs and campaigns that inform individuals about the impact of their food choices and encourage healthier and more sustainable diets can lead to positive change.

3.   **Supporting Local and Sustainable Food Systems:** Supporting local farmers and sustainable food practices can help address ethical concerns in food systems.   Farmers' markets, community-supported agriculture programs, and urban gardening initiatives can contribute to healthier, more equitable, and environmentally friendly food systems.

4. **Collaboration and Partnerships:** Addressing the complex web of ethical issues in food systems requires collaboration and partnerships across sectors. This can involve cooperation between governments, civil society organizations, farmers, and industry stakeholders. By working together, it is possible to develop comprehensive solutions that prioritize health, equity, and sustainability.

In conclusion, food systems have profound ethical implications for health and well-being. Ensuring access to safe, nutritious, and sustainable food is not only a matter of personal choice but also a societal responsibility. By addressing the ethical issues in food systems, we can promote healthier individuals, stronger communities, and a more equitable and sustainable world.

Urbanization and Health

Urbanization refers to the process of the population shifting from rural areas to urban areas, resulting in the growth and development of cities. This trend has been on the rise globally, with more people choosing to live in urban areas for various reasons such as employment opportunities, better access to healthcare and education, and improved quality of life. However, urbanization also brings with it a unique set of challenges and ethical considerations related to health and wellness.

Understanding the Impact of Urbanization on Health

Urbanization can have both positive and negative effects on health and wellness. On one hand, cities often offer better healthcare facilities, advanced medical technology, and specialized healthcare services compared to rural areas. Urban areas also tend to have higher standards of living, improved sanitation systems, and better access to nutritious food, which can contribute to better health outcomes.

On the other hand, rapid urbanization can lead to overcrowding, inadequate housing, and the emergence of slums or informal settlements. These conditions can increase the risk of infectious diseases, poor sanitation, and limited access to healthcare services for marginalized populations. Additionally, urban areas may also have higher levels of air pollution, noise pollution, and occupational hazards, which can have negative impacts on respiratory health and overall well-being.

Addressing Health Inequities in Urban Settings

Urbanization often exacerbates existing health inequities, with marginalized populations being disproportionately affected by the negative health consequences of urban living. Efforts should be made to address these inequities and ensure that urban development promotes health and well-being for all residents.

One approach is to adopt a health equity lens in urban planning and design. This involves considering the health impacts of decisions related to housing, transportation, and infrastructure. For example, ensuring access to affordable housing, green spaces, and public transportation systems can contribute to improved physical and mental health outcomes. This requires collaboration between urban planners, policymakers, healthcare professionals, and community members to prioritize health and equity in decision-making processes.

Promoting Active and Healthy Lifestyles

Urban environments can either facilitate or hinder healthy lifestyle choices. Access to safe and well-maintained sidewalks, parks, and recreational spaces can encourage physical activity and reduce the risk of chronic diseases such as obesity, diabetes, and cardiovascular diseases. On the contrary, limited access to such amenities or concerns about safety can discourage physical activity and contribute to sedentary lifestyles.

To promote active and healthy lifestyles in urban areas, it is important to design neighborhoods and public spaces that prioritize pedestrian and cyclist safety. This can include creating dedicated bike lanes, improving street lighting, and implementing traffic calming measures. Additionally, urban planning should prioritize the inclusion of parks, playgrounds, and community gardens to encourage outdoor activities and social interaction.

Mitigating Environmental Health Risks

Urbanization often leads to increased environmental health risks, including air pollution, water pollution, and exposure to hazardous substances. These risks can have profound health impacts, ranging from respiratory illnesses to increased risk of chronic diseases and environmental-related cancers.

To mitigate environmental health risks in urban areas, it is crucial to implement policies and interventions that address pollution sources. This can involve promoting the use of renewable energy sources, improving waste management systems, and strengthening environmental regulations. Additionally, individuals can play a role by adopting sustainable practices such as reducing personal energy consumption, practicing responsible waste disposal, and advocating for environmentally friendly policies.

Building Resilient Urban Health Systems

As urban populations continue to grow, it is essential to build resilient health systems that can effectively respond to the unique challenges of urban living. This includes ensuring accessibility, affordability, and quality of healthcare services for all residents.

One approach is to develop integrated and decentralized healthcare systems that are responsive to the diverse needs of urban populations. This can involve establishing primary healthcare centers in underserved areas, promoting community-based care models, and leveraging technology to enhance healthcare delivery, such as telemedicine and mobile health applications.

Furthermore, urban health systems should prioritize preventive care, health promotion, and early intervention. This includes implementing health education programs, promoting vaccination campaigns, and conducting regular screenings for common urban health concerns such as diabetes, cardiovascular diseases, and mental health disorders.

Case Study: The Health Impact of Urban Renewal Projects

Urban renewal projects, aimed at revitalizing deteriorating urban areas, can have significant health impacts on local communities. However, the ethical considerations of these projects often revolve around issues of displacement, gentrification, and inequitable distribution of resources.

For instance, the displacement of low-income residents due to rising property prices can disrupt community networks and lead to increased social isolation, stress, and mental health issues. Gentrification can also result in the loss of affordable housing, displacement of local businesses, and the erosion of cultural heritage.

To address these ethical concerns, urban renewal projects should prioritize inclusive development and community engagement. This involves involving local residents in decision-making processes, ensuring affordable housing options for existing residents, and preserving the cultural identity of the community. Additionally, comprehensive social support systems should be put in place to mitigate the potential negative health impacts of displacement and gentrification.

Conclusion

Urbanization presents a unique set of ethical considerations for health and wellness. While urban areas offer several advantages in terms of healthcare access and quality of life, rapid urbanization can also lead to health inequities, environmental health risks, and challenges in providing equitable healthcare services.   It is essential to adopt a holistic and collaborative approach that prioritizes health equity, promotes active and healthy lifestyles, mitigates environmental risks, and builds resilient urban health systems. By addressing these ethical considerations, we can strive towards a future where urbanization contributes to the health and well-being of all residents.

Ethical Issues in Healthcare Systems

Healthcare Access and Universal Coverage

The Ethics of Healthcare as a Human Right

Access to healthcare is a fundamental aspect of human well-being, and the ethics surrounding healthcare provision have profound implications for individuals and societies. In this section, we will explore the concept of healthcare as a human right and the ethical considerations that arise from this perspective.

Understanding Healthcare as a Human Right

The notion of healthcare as a human right is grounded in the belief that every individual has a fundamental entitlement to access necessary healthcare services without discrimination or financial hardship. This perspective is based on the principles of human dignity, equality, and social justice.

Health, as defined by the World Health Organization (WHO), is not merely the absence of disease or infirmity but a state of complete physical, mental, and social well-being. When we recognize that health is central to the overall quality of life and an essential component of human flourishing, the argument for healthcare as a human right becomes compelling.

The ethical framework that underlies the concept of healthcare as a human right is rooted in principles such as autonomy, justice, and beneficence.

Autonomy and Healthcare Choice

Autonomy, the principle of self-determination and decision-making, plays a crucial role in understanding healthcare as a human right. It acknowledges an individual's right to make choices that affect their own health and well-being.

From an ethical standpoint, respecting autonomy means ensuring that individuals have the freedom to access healthcare services that align with their values, preferences, and needs. This includes the right to make informed decisions about their treatment options, give or withhold consent, and have control over their own bodies.

However, autonomy does not exist in isolation; it must be balanced with other ethical principles to ensure its promotion does not lead to harm or injustice.

Justice and Healthcare Equity

The principle of justice is central to the discussion of healthcare as a human right. Justice involves the fair allocation of healthcare resources and aims to eliminate barriers to access based on factors such as income, social status, race, gender, or geographic location.

Achieving healthcare equity requires addressing existing disparities in healthcare access and outcomes. It involves adopting policies that promote equal opportunities for all individuals to obtain affordable, appropriate, and timely healthcare services.

Issues related to justice in healthcare include ensuring universal coverage, reducing socioeconomic disparities in health, and addressing systemic biases that disadvantage marginalized and vulnerable populations.

Beneficence and Healthcare Provision

The principle of beneficence, which emphasizes the duty to do good and promote well-being, is a guiding ethical principle in healthcare. It calls on healthcare providers and policymakers to act in ways that maximize the benefit and minimize harm to individuals and communities.

Applying the principle of beneficence to healthcare as a human right means ensuring that everyone has access to quality healthcare services that meet their needs. This includes preventive care, essential treatments, and long-term support for chronic conditions. It also requires addressing social determinants of health to promote overall well-being.

Healthcare providers have a professional responsibility to deliver care that is evidence-based, culturally sensitive, and respectful of patients' values and preferences. In doing so, they actively contribute to the realization of healthcare as a human right.

Challenges and Solutions

While the idea of healthcare as a human right is widely recognized, implementing and ensuring its realization can be challenging. Some of the key challenges include:

1. **Resource Allocation:** Healthcare systems often face limited resources, necessitating difficult decisions regarding the allocation of finite resources. Ethical frameworks like principles of distributive justice and priority-setting can guide fair resource allocation.

2. **Financial Constraints:** Funding healthcare for all individuals, particularly in low-income countries or economically disadvantaged communities, requires sustainable financing mechanisms and effective resource allocation strategies. Governments, international organizations, and civil society can collaborate to address financial constraints and ensure equitable access to healthcare.

3. **Healthcare Disparities:** Marginalized populations, such as racial and ethnic minorities, indigenous communities, and those living in rural or remote areas, often face greater barriers to access healthcare due to structural inequalities. Addressing healthcare disparities requires targeted interventions, community engagement, and policy changes.

4. **Public Health Emergencies:** During emergencies like pandemics or natural disasters, ensuring healthcare as a human right becomes even more critical. Ethical decision-making frameworks and crisis management strategies should prioritize the protection of vulnerable populations and allocate resources in a just and equitable manner.

Ultimately, recognizing healthcare as a human right requires a comprehensive, multi-dimensional approach that combines ethical principles, policy changes, resource allocation, and social determinants of health considerations. By addressing the challenges and embracing the opportunities, we can strive towards a more equitable and just healthcare system that respects the dignity of every individual.

Conclusion

The recognition of healthcare as a human right underscores the ethical imperative to promote access to quality healthcare services for all individuals. By embracing principles such as autonomy, justice, and beneficence, we can work towards an equitable healthcare system that values human dignity, eliminates disparities, and promotes overall well-being.

As we navigate the complexities of healthcare ethics, it is crucial to continuously engage in critical reflection, dialogue, and advocacy, ultimately shaping a future where health and wellness are truly universal human rights.

Further Reading

+ Gostin, L. O. (2019). Global health law. Harvard University Press.

+ Childress, J. F., & Bernheim, R. G. (2019). Beyond Therapy: Biotechnology and the Pursuit of Happiness. Cambridge University Press.

+ Powers, M., & Faden, R. (2006). Social justice: The moral foundations of public health and health policy. Oxford University Press.

Healthcare Reform and Equity

Healthcare reform is a vital and ongoing process that seeks to improve the accessibility, affordability, and quality of healthcare services. One of the central concerns of healthcare reform is equity, ensuring that all individuals have equal access to healthcare resources and that healthcare outcomes are not determined by socio-economic status, race, gender, or other factors. In this section, we will explore the ethical considerations surrounding healthcare reform and the pursuit of equity in healthcare systems.

The Need for Healthcare Reform

Before delving into the ethical considerations, it is important to understand the need for healthcare reform. In many countries, healthcare systems face various challenges and inequalities that hinder the provision of quality healthcare to all individuals. These challenges may include disparities in access to healthcare services, rising healthcare costs, inadequacy of health insurance coverage, and fragmented healthcare delivery systems.

Healthcare reform aims to address these challenges, and in doing so, improve health outcomes for individuals and communities. It seeks to eliminate barriers to healthcare access, improve the affordability of healthcare services, enhance the quality of care provided, and create a more efficient and sustainable healthcare system.

Ethical Principles in Healthcare Reform

To guide healthcare reform efforts, it is essential to consider key ethical principles that uphold equity and justice in healthcare systems. Several principles are particularly relevant in this context:

1. **Principle of Equality**: This principle asserts that all individuals should have equal access to healthcare services and resources. It requires eliminating any unjustifiable disparities in healthcare access and outcomes based on factors such as socio-economic status, race, ethnicity, gender, or geographic location.

2. **Principle of Solidarity**: Solidarity emphasizes the collective responsibility of society to ensure equitable access to healthcare. It calls for individuals and institutions to work together for the common good and to address health inequalities as a shared responsibility.

3. **Principle of Justice**: Justice involves fairness in the distribution of healthcare resources, ensuring that healthcare services are allocated based on need rather than ability to pay. It also encompasses the principle of proportionality, which asserts that the allocation of resources should be proportional to the expected health benefit.

4. **Principle of Autonomy**: Autonomy recognizes individuals' rights to make decisions about their own healthcare. Healthcare reform should respect and promote individual autonomy by providing individuals with information, choices, and the ability to make informed decisions about their healthcare.

These ethical principles provide a foundation for reforming healthcare systems with the aim of achieving equity and justice.

Key Strategies for Healthcare Reform

To achieve equity in healthcare, healthcare reform efforts must focus on implementing key strategies. Here are some strategies that have been widely discussed and implemented:

1. **Universal Healthcare Coverage**: Implementing a system of universal healthcare coverage ensures that all individuals have access to essential healthcare services without financial hardship. This can be achieved through public financing mechanisms, such as a single-payer system or social health insurance, which pool health risks and resources to provide comprehensive healthcare coverage.

2. **Healthcare Workforce Development:** Healthcare reform should prioritize the development and distribution of a well-trained and diverse healthcare workforce. This includes increasing the number of healthcare professionals in underserved areas, promoting cultural competence, and addressing workforce shortages.

3. **Investment in Prevention and Primary Care:** Shifting the focus from a predominantly acute care model to a preventive and primary care model can improve health outcomes and reduce healthcare costs. Investing in preventive measures, health promotion, and early intervention can help detect and address health issues before they escalate, leading to better overall health and reduced healthcare disparities.

4. **Eliminating Health Disparities:** Addressing and eliminating health disparities is crucial in healthcare reform. This involves understanding and addressing the social determinants of health that contribute to disparities, such as income inequality, racial discrimination, and inadequate access to education and employment opportunities.

5. **Health Information Technology:** Leveraging health information technology can improve healthcare delivery, coordination, and efficiency. Electronic health records, telemedicine, and telehealth services can enhance access to healthcare, particularly in underserved areas. However, it is important to ensure that these technologies are accessible to all and do not further exacerbate existing health disparities.

Ethical Challenges and Considerations

While healthcare reform aims to promote equity and justice, it also presents several ethical challenges. These challenges should be considered to develop effective and ethical healthcare reform policies:

1. **Resource Allocation:** Determining how healthcare resources should be allocated is a complex ethical issue. Healthcare reform must address questions such as how to prioritize certain treatments or services, how to distribute resources fairly across different populations, and how to balance the needs of individuals and the collective population.

2. **Rationing of Care:** In situations where resources are limited, difficult decisions may need to be made regarding the availability of certain treatments or services. Ethical considerations in healthcare reform involve

developing fair and transparent processes for rationing care, ensuring that decisions are evidence-based, unbiased, and uphold the principles of equity and justice.

3. **Public versus Private Healthcare:** Healthcare reform often involves debates about the role of the public and private sectors in the provision of healthcare. Balancing the benefits and drawbacks of each system requires careful consideration to ensure equitable access, affordability, and quality of care for all individuals.

4. **Stakeholder Engagement:** Healthcare reform must involve meaningful engagement and collaboration with a wide range of stakeholders, including healthcare professionals, patients, communities, policymakers, and advocacy groups. Ethical considerations include ensuring diverse representation, respecting the perspectives and values of all stakeholders, and creating spaces for open dialogue and shared decision-making.

Case Study: Affordable Care Act (ACA)

An example of healthcare reform aimed at improving equity and access is the Affordable Care Act (ACA) implemented in the United States. The ACA sought to expand health insurance coverage, protect individuals with pre-existing conditions, and promote preventive care.

However, the ACA faced ethical challenges, including controversies surrounding the individual mandate, concerns over rising healthcare costs, and limitations in achieving universal coverage. The ACA also highlighted the tensions between individual autonomy and collective responsibility in healthcare reform.

Despite its limitations, the ACA has contributed to reducing the uninsured rate and improving access to healthcare for many individuals. It serves as a valuable case study for understanding the complexities and ethical considerations involved in healthcare reform.

Conclusion

Healthcare reform and equity are central to creating a fair and just healthcare system that provides quality care to all individuals. Ethical considerations, guided by principles such as equality, solidarity, justice, and autonomy, should drive healthcare reform efforts.

By implementing strategies such as universal healthcare coverage, investing in prevention and primary care, and addressing health disparities, healthcare systems can become more equitable and improve health outcomes for all. However, ethical challenges such as resource allocation, rationing of care, and stakeholder engagement must be carefully navigated to ensure the ethical foundation of healthcare reform.

Ultimately, healthcare reform should strive to achieve a healthcare system that is characterized by accessibility, affordability, quality, and fairness, creating a healthier and more equitable future for all.

Global Disparities in Healthcare Access

One of the most pressing ethical issues in healthcare is the global disparity in access to healthcare services. While health is a fundamental human right, millions of people around the world do not have access to even the most basic healthcare services. This section explores the root causes of global disparities in healthcare access and examines potential ethical solutions to this critical problem.

Understanding Global Disparities in Healthcare Access

Definition and Key Concepts

Global disparities in healthcare access refer to the unequal distribution of healthcare resources, services, and opportunities among different countries and populations. Access to healthcare is influenced by various factors, including socioeconomic status, education, geographical location, and political stability. These disparities result in significant differences in health outcomes and life expectancy between countries and communities.

Historical Overview of Global Health Inequalities

The inequities in healthcare access that exist today are deeply rooted in historical and colonial legacies. During the colonial era, access to healthcare was often limited to the ruling elite, while indigenous populations were marginalized and excluded from these services. This historical injustice has contributed to the perpetuation of global health disparities.

Furthermore, the global economic system has perpetuated inequalities in healthcare access. Low-income countries often lack the financial resources and infrastructure to develop and maintain efficient healthcare systems, while high-income countries enjoy better access to healthcare services.

Intersectionality and Health Disparities

Global disparities in healthcare access are further compounded by intersectionality, which refers to the interconnected nature of various social identities and systems of oppression. Intersectionality recognizes that individuals experience privilege and oppression based on multiple factors such as race, class, gender, and ethnicity.

For example, women in many countries face additional barriers to accessing healthcare due to societal norms, patriarchy, and cultural practices. Similarly, individuals from minority racial or ethnic groups may encounter discrimination and bias when seeking healthcare services. Intersectionality highlights the need for a comprehensive and inclusive approach to addressing healthcare disparities.

Economic Inequalities and Healthcare Access

Economic inequalities play a significant role in global disparities in healthcare access. Poverty is a major barrier to receiving adequate healthcare, as individuals living in poverty often lack the financial means to access healthcare services, afford medications, or receive necessary treatments.

Moreover, the cost of healthcare services, such as hospital visits, medications, and surgeries, can be prohibitive for individuals in low-income countries. This economic burden leads to delayed or inadequate healthcare, exacerbating health conditions and increasing mortality rates.

Social Determinants of Health and Healthcare Access

Social determinants of health, such as education, income, employment, and social support, significantly influence healthcare access. Individuals with lower levels of education and income are more likely to face barriers in accessing healthcare services.

Inadequate education limits individuals' understanding of health issues and their ability to navigate the healthcare system. Unemployment and low-income levels restrict financial resources available for healthcare. Lack of social support networks can lead to isolation and limited access to healthcare information and services.

Ethical Considerations in Addressing Global Disparities in Healthcare Access

Addressing global disparities in healthcare access requires ethical considerations that prioritize justice, equity, and human rights. The following ethical principles can guide the development of strategies and policies for global health equity:

1. **Equity and Justice:** Every individual has a right to equitable access to healthcare services regardless of their socioeconomic status, geographic location, or demographic characteristics. Achieving health equity requires addressing the root causes of disparities and ensuring that resources are distributed fairly.

2. **Solidarity:** Global health is a collective responsibility, and solidarity emphasizes the need for collaboration and cooperation among nations to address healthcare disparities. It involves sharing resources, knowledge, and expertise to promote equitable access to healthcare.

3. **Human Rights:** Access to healthcare is recognized as a fundamental human right. The Universal Declaration of Human Rights affirms the right to the highest attainable standard of physical and mental health. International organizations like the World Health Organization (WHO) play a crucial role in advocating for the protection of this right.

4. **Social Determinants of Health:** Recognizing the impact of social determinants of health on healthcare access is essential. Addressing broader social and economic factors, such as education, income, and social support, can help reduce disparities and improve access to healthcare services.

Solutions and Approaches to Addressing Healthcare Disparities

Effectively addressing global disparities in healthcare access requires a comprehensive and collaborative approach. Here are some potential solutions and approaches to consider:

1. **Strengthening Healthcare Systems:** Investing in the development and strengthening of healthcare systems in low-income countries is crucial. This includes improving infrastructure, healthcare workforce capacity, and access to essential medicines and technologies.

2. **International Cooperation and Aid:** High-income countries can provide financial and technical support to low-income countries to enhance their healthcare systems. International organizations and partnerships can facilitate cooperation and ensure equitable distribution of resources.

3. **Education and Awareness:** Increasing healthcare literacy and awareness among communities can empower individuals to demand and navigate healthcare services effectively. Education programs can focus on preventive care, health promotion, and addressing misconceptions about healthcare.

4. **Health Workforce Training:** Providing training and education for healthcare professionals in culturally competent care, addressing implicit bias, and promoting diversity and inclusion can help reduce healthcare disparities related to discrimination and stigma.

5. **Innovative Approaches:** Embracing technological advancements, such as telemedicine and mobile health applications, can help overcome geographical barriers and improve access to healthcare in underserved areas.

Real-World Example: Partners in Health

Partners In Health (PIH) is an example of an organization that addresses global disparities in healthcare access. Founded in 1987, PIH works in partnership with local communities and governments to provide high-quality healthcare services and build sustainable healthcare systems in resource-limited settings.

PIH takes a comprehensive approach, addressing social determinants of health, such as poverty, education, and housing, alongside medical interventions. They prioritize building long-term relationships and ensuring the dignity and respect of patients. PIH's approach emphasizes community engagement, training and empowering local healthcare workers, and advocating for policy changes to improve healthcare access for all.

Through their work, PIH demonstrates how ethical principles, such as equity, solidarity, and a patient-centered approach, can guide efforts to reduce global disparities in healthcare access.

Summary

Global disparities in healthcare access represent a significant challenge that requires ethical solutions. Understanding the root causes, such as economic inequalities, social determinants of health, and historical legacies, is crucial to developing effective strategies. Ethical considerations, including justice, equity, human rights, and solidarity, play a vital role in addressing these disparities and striving for a more equitable and just healthcare future.

By strengthening healthcare systems, promoting international cooperation, investing in education and awareness, training healthcare professionals, and embracing innovation, we can work towards reducing global disparities in healthcare access and ensuring that everyone has the opportunity to live a healthy and fulfilling life.

Healthcare Allocation and Resource Allocation

Rationing and Priority Setting

In the healthcare system, the allocation of resources is often a challenging task. When resources are limited, difficult decisions must be made regarding who receives what healthcare services and treatments. Rationing and priority setting are ethical considerations in healthcare that aim to address this issue. In this section, we will explore the principles and frameworks used in rationing and priority setting, examine the challenges involved, and discuss potential solutions.

Principles of Rationing

Rationing refers to the process of distributing scarce resources in a fair and equitable manner. When resources such as organs for transplantation, expensive medications, or intensive care unit (ICU) beds are limited, rationing becomes necessary. Several ethical principles guide the rationing process:

1. **Justice:** The principle of justice asserts that healthcare resources should be distributed fairly and equitably, ensuring equal opportunity for all individuals in need. However, defining what constitutes fairness can be challenging, as different people may have varying opinions on what is fair.

2. **Utility:** The principle of utility suggests that resources should be allocated in a way that maximizes overall societal benefit. This principle prioritizes the allocation of resources to individuals who can derive the greatest benefit from them. However, determining the level of benefit and how to measure it objectively can be complex.

3. **Priority to the Worst Off:** This principle advocates for prioritizing individuals who are in the worst health condition or have the greatest health needs. It aims to address health disparities and ensure that those who need healthcare the most receive it. However, determining who qualifies as the worst off can be subjective and may raise questions of fairness.

4. **Proportionality:** The principle of proportionality suggests that the allocation of resources should be proportional to the individual's contribution to society or their responsibility for the condition requiring healthcare. This principle considers factors such as personal choices, behaviors, or social contributions when making allocation decisions. However, applying this principle can be ethically contentious and challenging.

5. **Transparency and Accountability:** The principle of transparency emphasizes the importance of open and honest communication about the rationing process. It involves clear guidelines for decision-making, providing explanations for resource allocation decisions, and allowing for public scrutiny. Accountability ensures that those responsible for making allocation decisions are held responsible for their choices.

These principles provide a framework for making fair and ethical decisions regarding resource allocation in healthcare. However, applying them in practice is not without challenges.

Challenges in Rationing

Rationing healthcare resources is fraught with complexities and ethical dilemmas. Some of the key challenges in implementing a fair and effective rationing system include:

+ **Limited Resources:** The scarcity of resources poses a significant challenge in rationing decisions. Deciding who should receive a limited number of organs for transplantation or deciding which patients should occupy ICU beds can be emotionally and ethically challenging.

+ **Subjectivity and Variability:** The criteria used to determine priority in resource allocation can be subjective and vary across different healthcare systems and cultures. This subjectivity can raise questions of fairness and equity.

+ **Public Perception and Trust:** Rationing decisions may face public scrutiny and criticism. It is essential to maintain public trust by ensuring transparency, involving stakeholders in decision-making processes, and providing clear explanations for resource allocation choices.

+ **Time Constraints:** In urgent situations, such as during a pandemic or a disaster, there may be limited time for thorough decision-making processes.

This can make it challenging to ensure fairness and gather input from various stakeholders.

+ **Legal and Regulatory Implications:** Rationing decisions may have legal and regulatory implications. Balancing ethical principles with existing laws and regulations can be complex and require careful consideration.

Addressing these challenges calls for innovative solutions and effective frameworks for rationing and priority setting in healthcare.

Solutions and Frameworks

To navigate the challenges of rationing and priority setting, healthcare systems employ various frameworks and strategies. Some of these include:

+ **Evidence-Based Guidelines:** Developing evidence-based guidelines can help standardize decision-making processes and reduce variability. These guidelines are based on rigorous research and aim to guide healthcare professionals in allocating resources effectively.

+ **Ethics Committees:** Ethics committees, comprised of diverse stakeholders, including healthcare professionals, ethicists, and members of the community, can provide guidance and support in difficult rationing decisions. These committees can help ensure that decisions are fair, transparent, and consistent with ethical principles.

+ **Public Engagement and Deliberation:** Involving the public in the decision-making process can increase transparency and build public trust. Public engagement methods, such as citizen juries or deliberative forums, allow for meaningful discussion and input from the community.

+ **Decision-Making Tools:** Decision-making tools, such as decision trees or scoring systems, can aid in systematically assessing individuals' eligibility for specific healthcare resources. These tools can help introduce objectivity into the process and minimize bias.

+ **Regular Review and Evaluation:** Regular review and evaluation of the rationing process and its outcomes are crucial for identifying areas of improvement, ensuring fairness, and adapting to changing healthcare needs.

Innovative solutions and ongoing research are essential to continually refine and improve the process of rationing and priority setting in healthcare. By integrating ethical principles, involving stakeholders, and promoting transparency, healthcare systems can strive to allocate resources in a fair and effective manner.

Real-World Example: Allocation of ICU Beds During the COVID-19 Pandemic

The COVID-19 pandemic created an unprecedented demand for ICU beds worldwide. Healthcare systems were faced with the challenge of allocating limited ICU beds to an overwhelming number of critically ill patients. Many countries developed triage protocols and guidelines to address this crisis.

For example, the Sequential Organ Failure Assessment (SOFA) score, which quantifies organ dysfunction, was used as a tool for determining priority in ICU admissions. Patients with higher SOFA scores, indicating more severe organ dysfunction, were given priority for ICU beds. This approach aimed to allocate resources to those with a higher likelihood of survival.

However, implementing such protocols raised ethical concerns. Critics argued that the SOFA score might disadvantage certain patient populations, such as the elderly or those with pre-existing conditions, who may have lower scores due to age or comorbidities. This raised questions about the fairness and equity of resource allocation.

Addressing these concerns required ongoing evaluation, feedback, and adaptation of the protocols. Healthcare systems had to balance the need for objective criteria with a sensitivity to individual circumstances. Public engagement and clear communication about the rationale for these decisions were essential for maintaining trust during this challenging time.

Key Takeaways

Rationing and priority setting in healthcare involve making difficult decisions about resource allocation in the face of limited resources. Ethical principles, including justice, utility, prioritizing the worst off, proportionality, transparency, and accountability, guide the allocation process.

Implementing a fair and effective rationing system comes with challenges, such as limited resources, subjectivity, public perception, time constraints, and legal implications. To address these challenges, healthcare systems use evidence-based guidelines, ethics committees, public engagement, decision-making tools, and regular review and evaluation.

Real-world examples, such as the allocation of ICU beds during the COVID-19 pandemic, highlight the complexities and ethical dilemmas associated with rationing decisions.

By balancing ethical principles, involving stakeholders, and promoting transparency, healthcare systems can strive towards a more equitable and just allocation of resources, ensuring that those most in need receive the care they require.

The Role of Cost-effectiveness in Healthcare Decision-making

In healthcare, decision-making can be complex, with various factors to consider, such as effectiveness, safety, and cost. Cost-effectiveness analysis (CEA) is a valuable tool used to assess the economic efficiency of healthcare interventions. It enables decision-makers to compare the costs and outcomes of different interventions, helping them determine the most cost-effective options. In this section, we will explore the role of cost-effectiveness in healthcare decision-making, its principles, methods, and ethical considerations.

Principles of Cost-effectiveness Analysis

CEA is based on the principle of maximizing health outcomes given a limited budget. It involves comparing the costs and outcomes of alternative interventions, typically expressed in a common unit, such as quality-adjusted life years (QALYs) gained. The key principles of CEA include:

1. **Comparative Analysis:** CEA involves comparing multiple interventions to determine their relative cost-effectiveness. This helps decision-makers identify the most efficient allocation of resources.

2. **Incremental Analysis:** CEA focuses on assessing the additional costs and outcomes achieved by adopting a specific intervention compared to a baseline or comparator intervention. This incremental approach helps evaluate the value of the additional benefits gained.

3. **Social Perspective:** CEA considers the societal perspective by including all relevant costs and outcomes, regardless of who incurs them. It takes into account direct medical costs, as well as indirect costs, such as productivity losses.

4. **Time Horizon:** CEA considers both short-term and long-term effects of interventions. It examines the costs and outcomes over the lifetime of the intervention, capturing potential future benefits and costs.

Methods of Cost-effectiveness Analysis

Several methods are commonly used in CEA to assess the cost-effectiveness of healthcare interventions. These methods provide decision-makers with valuable insights into the value and efficiency of different interventions. Some commonly used methods include:

1. **Cost-Effectiveness Ratio (CER):** The CER compares the costs of an intervention to its health outcomes. It is calculated by dividing the difference in costs between two interventions by the difference in their outcomes. The CER provides a simple measure of cost-effectiveness.

2. **Incremental Cost-Effectiveness Ratio (ICER):** The ICER compares the additional costs of implementing a new intervention to its additional health outcomes compared to a comparator intervention. It is calculated by dividing the incremental costs by the incremental outcomes. The ICER helps decision-makers assess the value of adopting a new intervention.

3. **Cost-Effectiveness Acceptability Curve (CEAC):** The CEAC provides a visual representation of the uncertainty surrounding the cost-effectiveness of an intervention. It shows the probability of an intervention being cost-effective at different willingness-to-pay thresholds. The CEAC helps decision-makers assess the robustness of the cost-effectiveness results.

Ethical Considerations in Cost-effectiveness Analysis

While cost-effectiveness analysis is a useful tool for guiding healthcare decision-making, it also raises ethical considerations. It is crucial to consider equity, fairness, and distributive justice when assessing the cost-effectiveness of healthcare interventions. Some key ethical considerations in CEA include:

1. **Equitable Allocation of Resources:** CEA should consider the impact of interventions on different populations and ensure equitable distribution of resources. Decision-makers must consider the potential disparities and strive to prioritize interventions that address the healthcare needs of disadvantaged populations.

2. **Value Pluralism:** CEA involves assigning values and priorities to different health outcomes. It is essential to recognize and take into account the diversity of values and preferences within society when evaluating the cost-effectiveness of interventions.

3. **Transparency and Stakeholder Engagement:** CEA should be conducted with transparency, involving various stakeholders, including patients, healthcare providers, and the public. Transparent decision-making processes help build trust, acceptability, and legitimacy.

4. **Prudence in Resource Allocation:** While cost-effectiveness analysis helps identify efficient resource allocation, it is essential to exercise prudence in decision-making. Not all valuable interventions may be cost-effective, and other ethical considerations, such as human rights and dignity, should also be taken into account.

Case Study: Cost-effectiveness Analysis of Vaccination Programs

To illustrate the role of cost-effectiveness analysis in healthcare decision-making, let's consider a case study on vaccination programs. Vaccination is a cost-effective public health intervention that prevents the spread of infectious diseases and reduces healthcare costs. Cost-effectiveness analysis can help guide decisions related to vaccine implementation.

Suppose we compare two vaccination programs for a particular disease: Program A and Program B. Program A is a one-time vaccination administered at a cost of $50 per individual, while Program B is a two-dose vaccination administered at a cost of $70 per individual.

To evaluate the cost-effectiveness, we consider the effectiveness of each program. Program A prevents 80% of infections, while Program B prevents 90% of infections. Additionally, we consider the costs of treating an infected individual, including hospitalization, medication, and lost productivity.

Using cost-effectiveness analysis, we can calculate the ICER, which represents the additional cost per QALY gained for Program B compared to Program A. Let's assume that Program B provides an additional 0.1 QALYs per individual compared to Program A.

$$\text{ICER} = \frac{\text{Cost of Program B} - \text{Cost of Program A}}{\text{QALYs gained with Program B} - \text{QALYs gained with Program A}}$$

Substituting the values:

$$\text{ICER} = \frac{\$70 - \$50}{0.1 \ \text{QALYs}}$$

$$\text{ICER} = \$200$$

The ICER of $200 per QALY gained indicates that Program B is cost-effective compared to Program A, as it falls within the accepted cost-effectiveness threshold (usually set by health authorities).

However, ethical considerations should also guide decision-making in vaccine implementation. These considerations may include equitable access to vaccines, protection of vulnerable populations, and potential public health benefits beyond individual outcomes.

In conclusion, cost-effectiveness analysis plays a crucial role in healthcare decision-making by providing insights into the economic efficiency of different interventions. It helps allocate resources wisely, considering value for money and maximizing health outcomes. However, ethical considerations, such as equity and transparency, must be integrated into the decision-making process to ensure fairness and just distribution of resources.

Ethical Considerations in End-of-life Care

End-of-life care is a sensitive and complex topic that involves addressing the medical, emotional, and ethical needs of individuals nearing the end of their lives. In this section, we will examine the ethical considerations that arise in the context of end-of-life care, including issues related to decision-making, the withdrawal of treatment, and physician-assisted death. We will also explore the importance of palliative care and the role of healthcare professionals in providing compassionate and ethical care to patients at the end of life.

Autonomy and Decision-making

One of the fundamental ethical principles in end-of-life care is respect for patient autonomy. Autonomy refers to an individual's right to make decisions about their own healthcare, including decisions about end-of-life care. It is essential to ensure that individuals have the capacity to make informed decisions and are provided with the necessary information and support to exercise their autonomy.

However, decision-making in end-of-life care can be complex, especially when patients are unable to communicate their wishes. In such cases, advance care planning becomes crucial. Advance care planning involves discussions between healthcare professionals, patients, and their families to create a plan that reflects the patient's values, goals, and preferences for end-of-life care. This plan may include the designation of a healthcare proxy or the creation of a living will, which outlines the patient's wishes regarding life-sustaining treatments.

Healthcare professionals must respect and honor these advance care plans, ensuring that decisions align with the patient's expressed wishes. However, it is essential to recognize that advance directives may not cover all possible scenarios, and thus healthcare professionals should engage in ongoing dialogue with patients and their families to provide collaborative decision-making when unexpected situations arise.

Withdrawal of Treatment

Another ethical consideration in end-of-life care is the withdrawal of treatment. In some situations, the continuation of life-sustaining treatment may be futile or burdensome, and may not align with the patient's goals or values. In such cases, healthcare professionals may consider withdrawing or withholding treatment.

The decision to withdraw treatment must be made collaboratively, involving the patient (if possible) or the patient's designated healthcare proxy, and a multidisciplinary team of healthcare professionals. Ethical principles such as beneficence (doing what is in the best interest of the patient) and non-maleficence (avoiding harm to the patient) must guide the decision-making process.

Careful consideration should be given to ensuring that patients are comfortable and receive appropriate palliative care during the withdrawal of treatment. Palliative care focuses on managing the physical, emotional, and spiritual symptoms experienced by patients at the end of life, providing comfort and support to both the patient and their loved ones.

Physician-Assisted Death

Physician-assisted death, in which a physician provides the means for a patient to end their own life, is a contentious topic in end-of-life care. Different jurisdictions have different laws and regulations regarding physician-assisted death, and these laws are constantly evolving.

Ethical debates surrounding physician-assisted death often revolve around the concepts of autonomy, beneficence, and non-maleficence. Proponents argue that it allows patients to maintain control over their own lives and avoid unnecessary suffering, while opponents argue that it goes against the principles of the medical profession and may lead to unethical practices.

In jurisdictions where physician-assisted death is legal, stringent safeguards are typically in place to ensure that the decision-making process is thorough and robust. These safeguards often include requirements such as a terminal prognosis, the presence of unbearable suffering, multiple requests for assistance in dying, and the involvement of multiple healthcare professionals.

It is crucial for healthcare professionals to approach discussions about physician-assisted death with sensitivity and compassion, respecting the autonomy and dignity of each patient. It is also essential to provide alternative options, such as palliative care, to ensure that patients have access to comprehensive end-of-life care.

The Role of Healthcare Professionals

Healthcare professionals have a significant role to play in providing ethical and compassionate care to patients at the end of life. This includes not only providing medical care but also addressing the emotional, psychological, and spiritual needs of patients and their families.

Cultural competence is vital in end-of-life care, as patients may have diverse cultural and religious beliefs that influence their preferences and decisions. Healthcare professionals should strive to understand and respect these beliefs, tailoring care to meet each patient's unique needs.

Communication skills are also crucial in end-of-life care, as healthcare professionals need to have open and honest discussions with patients and their families about prognosis, treatment options, and end-of-life decisions. These discussions should be conducted in a compassionate and empathetic manner, allowing patients and their families to ask questions, express their concerns, and make informed decisions.

Healthcare professionals should also be aware of their own values and biases and strive to provide care that is free from judgment and discrimination. Allowing patients to guide the decision-making process and respecting their autonomy is essential in promoting ethical and patient-centered end-of-life care.

Conclusion

Ethical considerations in end-of-life care are complex and require healthcare professionals to navigate sensitive and emotionally charged situations. Respect for patient autonomy, collaboration in decision-making, and the provision of compassionate and comprehensive care are essential elements in promoting ethical and just end-of-life care. Palliative care plays a crucial role in managing symptoms and providing support to patients and their loved ones. By integrating these ethical principles into their practice, healthcare professionals can ensure that patients receive the care and support they need during this challenging phase of life.

Healthcare Professionalism and Responsibilities

Ethical Challenges in Healthcare Professional-Patient Relationships

In the field of healthcare, the relationship between healthcare professionals and patients is of utmost importance. This relationship forms the foundation of ethical and effective healthcare delivery. However, it is not without its challenges. In this section, we will explore some of the ethical challenges that can arise in healthcare professional-patient relationships and discuss strategies to address them.

Power Imbalance and Autonomy

One ethical challenge in healthcare professional-patient relationships relates to the power imbalance between the two parties. Healthcare professionals, due to their expertise and authority, often have more power and influence over patients' healthcare decisions. This power dynamic can potentially undermine patient autonomy, which refers to the right of patients to make informed decisions about their healthcare.

To address this challenge, healthcare professionals must respect and promote patient autonomy by ensuring that patients have access to accurate and unbiased information. This includes providing clear explanations of diagnoses, treatment options, and potential risks and benefits. Healthcare professionals should engage in shared decision-making with patients, considering their values, preferences, and unique circumstances. By doing so, they empower patients to actively participate in their own healthcare decisions and respect their autonomy.

Privacy and Confidentiality

Privacy and confidentiality are other critical ethical considerations in healthcare professional-patient relationships. Patients must trust that their personal health information will be kept confidential and that their privacy will be respected. However, in today's digital age, maintaining complete privacy and confidentiality can be challenging.

Healthcare professionals must adhere to strict ethical and legal standards to protect patient privacy and confidentiality. They should explain privacy policies, seek patient consent before sharing information, and take appropriate measures to safeguard electronic health records. Additionally, healthcare professionals should educate patients about their rights regarding their health information and involve them in decisions related to the sharing and storage of their data.

Cultural Competence and Communication

Cultural competence is another significant ethical challenge in healthcare professional-patient relationships, particularly in diverse societies. It involves understanding and respecting the cultural beliefs, values, and practices of patients. Effective communication is a vital component of cultural competence.

Healthcare professionals should strive to develop cultural competence by learning about the diverse cultural backgrounds of their patients. This includes understanding their beliefs about health and illness, language preferences, and healthcare practices. By doing so, healthcare professionals can tailor their communication and care plans to meet the individual needs of their patients.

Interpretation services and cultural mediators can also play a crucial role in bridging communication gaps and facilitating understanding between healthcare professionals and patients from different cultural backgrounds. Including patients' families or trusted individuals in discussions and decision-making processes can also enhance cultural competence and contribute to more ethical healthcare professional-patient relationships.

Boundary Issues

Maintaining appropriate boundaries in healthcare professional-patient relationships is another ethical challenge. Boundaries refer to the emotional, physical, and professional limits that should exist between healthcare professionals and patients. Violating these boundaries can harm the therapeutic relationship and compromise patient care.

Healthcare professionals should establish clear boundaries by maintaining professionalism and avoiding dual relationships or conflicts of interest. Dual relationships occur when healthcare professionals have personal, financial, or other non-professional relationships with their patients, which can create conflicts of interest. They should also be mindful of their behavior to ensure that it does not cross ethical boundaries, such as inappropriate touching, personal disclosure, or any form of exploitation.

Understanding the nuances of boundary issues and continuously reflecting on professional conduct can help healthcare professionals navigate this ethical challenge effectively.

Ethics of Care and Advocacy

In healthcare professional-patient relationships, ethical challenges can arise when healthcare professionals struggle with the tension between providing care and advocating for the best interests of their patients. This challenge can manifest in various ways, such as conflicting treatment options or limited resources.

Healthcare professionals must prioritize the well-being of their patients and act as their advocates. This involves being knowledgeable about patient rights and healthcare policies, advocating for equitable access to healthcare resources, and voicing concerns about potential ethical violations or disparities in healthcare delivery.

It is crucial for healthcare professionals to engage in ethical reflection and seek support from colleagues and professional organizations to navigate challenging situations and ensure that patient care remains paramount.

In conclusion, healthcare professional-patient relationships are inherently complex and can present various ethical challenges. By addressing power imbalances, respecting patient autonomy, safeguarding privacy and confidentiality, promoting cultural competence, maintaining appropriate boundaries, and prioritizing patient care and advocacy, healthcare professionals can navigate these challenges while upholding ethical principles in their practice. By doing so, they contribute to the development of a more ethical and just healthcare system.

Professional Autonomy and Conscientious Objection

Professional autonomy is a fundamental principle in healthcare ethics that recognizes the authority and independence of healthcare professionals to make decisions based on their expertise and judgment. It is rooted in the belief that professionals have the right to determine their own actions and the responsibility to act in the best interest of their patients.

Conscientious objection refers to the practice of healthcare professionals refusing to provide or participate in certain medical procedures or interventions based on their deeply held moral or religious beliefs. This may include procedures such as abortion, euthanasia, or certain forms of contraception. It is important to note that conscientious objection is not limited to healthcare professionals; other individuals in the healthcare system, such as pharmacists or nurses, may also raise objections.

While professional autonomy and conscientious objection are both important ethical concepts, they can sometimes come into tension with each other. The challenge lies in striking an appropriate balance between respecting healthcare professionals' autonomy and ensuring that patients receive the care they need.

Principles and Challenges

At the core of professional autonomy is the principle of respect for persons, which recognizes the dignity and autonomy of both healthcare professionals and patients. Healthcare professionals are individuals with their own beliefs, values, and moral compasses. They have the right to act in accordance with their conscience and not be compelled to provide or participate in procedures that conflict with their deeply held beliefs.

However, this principle must be balanced with other important ethical considerations, such as patient autonomy, beneficence, and justice. Patients have the right to receive appropriate and timely care, and healthcare professionals have a duty to provide care in a non-discriminatory manner. The challenge arises when a healthcare professional's conscientious objection potentially jeopardizes a patient's access to care or infringes upon their rights.

Navigating the Tension

To navigate the tension between professional autonomy and conscientious objection, it is necessary to establish clear guidelines and protocols that uphold the rights of both patients and healthcare professionals. Several strategies can be employed to address this issue effectively:

1. **Respectful Communication and Collaboration:** Effective and respectful communication between healthcare professionals and patients is essential to understand and address any potential conflicts. Open dialogue can promote understanding, shared decision-making, and ultimately, the provision of appropriate care.

2. **Providing Information and Alternative Options:** Healthcare professionals should ensure that patients have access to accurate and comprehensive information about their healthcare options. If a healthcare professional objects to a particular intervention, they should inform the patient about alternative providers or resources that can meet their needs.

3. **Establishing Institutional Policies:** Healthcare institutions should develop policies and guidelines that strike a reasonable balance between professional autonomy and patient rights. These policies should outline the process for conscientious objection, including provisions for timely referral or transfer of care when necessary.

4. **Training and Education:** Continuous education and training programs can help healthcare professionals develop skills in ethical decision-making, communication, and cultural sensitivity. By enhancing these competencies, professionals can navigate complex ethical situations while minimizing adverse effects on patient care.

5. **Legal Considerations:** Legislation around conscientious objection varies across jurisdictions. It is crucial for healthcare professionals and institutions to be familiar with local laws and regulations to ensure compliance and to protect the rights of both patients and professionals.

It is important to recognize that while conscientious objection is a legitimate expression of personal beliefs, it cannot be used as a justification for discrimination or denial of care. Healthcare professionals have an ethical obligation to provide appropriate care or to make timely and appropriate referrals in such cases.

Case Study: Medical Abortions and Conscientious Objection

Consider the case of Dr. Smith, an obstetrician-gynecologist who has a conscientious objection to performing medical abortions. A patient, Jane, visits Dr. Smith seeking a medical abortion. Dr. Smith believes that ending a pregnancy goes against their moral beliefs.

To navigate this situation, Dr. Smith should:

+ Engage in open and respectful communication with Jane to understand her needs, concerns, and values.

+ Provide Jane with accurate and unbiased information about her options, including medical abortion, and inform her about other healthcare professionals who can provide the desired service.

+ Offer appropriate referrals to healthcare professionals who are willing and able to provide medical abortions promptly.

+ Ensure continuity of care by collaborating with other healthcare professionals involved in Jane's care, while respecting patient confidentiality.

By following these steps, Dr. Smith upholds their professional autonomy while ensuring that Jane's autonomy and access to care are respected.

Resources for Further Exploration

+ American Medical Association (AMA). (2007). Opinion 1.1.7 - Physicians' Refusal to Provide Services Because of Conscientious Objection. Retrieved from `https://www.ama-assn.org/delivering-care/ethics/physicians-refusal-provide-services-because-conscientious-ob`

+ Royal College of Obstetricians and Gynaecologists (RCOG). (2018). The Role of Conscientious Objection in the Provision of Abortion Care. Retrieved from `https://www.rcog.org.uk/globalassets/documents/guidelines/statement/green-top-statement-ethical-ethical-consent-web.pdf`

+ General Medical Council (GMC). (2021). Personal Beliefs and Medical Practice. Retrieved from `https://www.gmc-uk.org/ethical-guidance/ethical-committee-guidance/personal-beliefs-and-medical-practice`

Conclusion

Professional autonomy and conscientious objection are complex ethical issues in healthcare. While healthcare professionals have the right to follow their conscience, it is crucial to strike a balance that respects patient autonomy, upholds access to appropriate care, and fosters a healthcare system that is equitable and just.

By promoting respectful communication, developing institutional policies, and providing education and training, healthcare professionals and institutions can navigate these tensions thoughtfully and ensure that the needs and rights of both healthcare professionals and patients are respected. In doing so, we move closer to achieving a healthcare system that embodies ethical principles and promotes the well-being of all.

Integrity and Professional Ethics

Integrity and professional ethics are essential aspects of healthcare that ensure the trust and confidence of patients and the public in healthcare professionals. As healthcare professionals, individuals have a moral and ethical responsibility to uphold the highest standards of integrity and ethical conduct in their practice. This section will explore the importance of integrity in healthcare, the principles underlying professional ethics, and the challenges related to integrity and professional ethics in healthcare.

The Importance of Integrity in Healthcare

Integrity is the adherence to moral and ethical principles, encompassing honesty, truthfulness, and consistency in actions and behaviors. In healthcare, integrity is crucial as it establishes the foundation for trust between healthcare professionals, patients, and society as a whole.

Healthcare professionals with integrity prioritize the welfare and well-being of their patients above personal gain or interests. They provide accurate information, make ethical decisions, respect patient autonomy, and maintain confidentiality. By demonstrating integrity, healthcare professionals build trust, foster open communication, and promote patient-centered care.

Principles of Professional Ethics

Professional ethics in healthcare are grounded in core principles that guide the conduct of healthcare professionals. These principles serve as a framework to ensure the provision of ethical and quality care. Some of the key principles include:

1. **Beneficence:** Healthcare professionals have a duty to act in the best interest of their patients, promoting their well-being and health outcomes.

2. **Non-maleficence:** Healthcare professionals must avoid causing harm to patients and refrain from actions that may result in harm.

3. **Autonomy:** Respecting patient autonomy involves recognizing and respecting their right to make decisions regarding their own healthcare.

Healthcare professionals should provide relevant information and involve patients in their treatment decisions.

4. **Justice:** Fairness in healthcare delivery is essential. Healthcare professionals must distribute healthcare resources equitably and treat all patients without discrimination.

5. **Veracity:** Healthcare professionals have an ethical obligation to be truthful and honest with their patients, providing accurate information and avoiding deception.

6. **Confidentiality:** Respecting patient privacy and maintaining confidentiality is vital to preserve trust between healthcare professionals and patients.

These principles provide a moral compass for healthcare professionals to navigate ethical dilemmas and make informed decisions in the best interest of their patients.

Challenges to Integrity and Professional Ethics in Healthcare

While integrity and professional ethics are fundamental principles in healthcare, there are various challenges that healthcare professionals may face in upholding these ideals. Some of these challenges include:

1. **Conflicts of Interest:** Healthcare professionals may encounter conflicts of interest that compromise their ethical decision-making. Conflicts between personal interests, financial incentives, and patient well-being can pose challenges to maintaining integrity.

2. **Boundary Violations:** Establishing and maintaining appropriate boundaries with patients is crucial. Healthcare professionals must be mindful of their relationships with patients to prevent boundary violations that can erode trust and compromise professional ethics.

3. **Workplace Pressures:** Healthcare professionals may experience workplace pressures that can impact their ethical conduct. Productivity demands, resource constraints, and organizational cultures may create dilemmas, requiring healthcare professionals to make ethical judgments in challenging situations.

4. **Inadequate Training:** Insufficient education and training in ethical decision-making can hinder healthcare professionals' ability to navigate complex ethical issues. Continuous professional development and comprehensive training in professional ethics are essential to ensure knowledge and skill development.

5. **Cultural and Social Factors:** Healthcare professionals work in diverse cultural and social contexts. Adhering to professional ethics requires awareness of cultural values, norms, and beliefs that may influence patient care and decision-making. Cultural competence is necessary to provide ethical and equitable care.

Addressing these challenges requires proactive efforts, including ongoing education and support systems within healthcare organizations to promote ethical practices and provide guidance to healthcare professionals when faced with ethical dilemmas.

Case Study: Addressing Conflicts of Interest

Conflicts of interest are common ethical challenges in healthcare. Consider the following scenario:

Dr. Smith is a physician who recently started their clinical practice in a small community. They discover that one of the largest pharmaceutical companies has offered them a substantial financial incentive to prescribe their newly developed medication to patients. Dr. Smith knows that alternative treatments may be equally effective but not financially supported by the pharmaceutical company.

Question: What steps can Dr. Smith take to address the conflict of interest and uphold integrity in their practice?

Solution: Dr. Smith should consider the impact of accepting the financial incentive on their ability to maintain patient-centered care. They should critically evaluate the potential bias that may arise from financial interests. Seeking guidance from colleagues, discussing the case with an ethics committee, or consulting professional codes of conduct can provide valuable insights to navigate the conflict of interest. Ultimately, prioritizing patient well-being and making evidence-based decisions in the absence of financial biases are essential to uphold integrity and professional ethics.

Conclusion

Integrity and professional ethics are fundamental in healthcare and play a vital role in ensuring patient trust, safety, and quality care. Upholding ethical principles, demonstrating integrity, and effectively navigating ethical challenges contribute to the establishment of a just and ethical healthcare system. By embracing responsibility and actively addressing ethical concerns, healthcare professionals can contribute to a sustainable and equitable future for health and wellness.

Cultural Competence and Healthcare Delivery

Cultural competence is a critical aspect of healthcare delivery that promotes equitable and inclusive care for individuals from diverse cultural backgrounds. It involves understanding and respecting the values, beliefs, practices, and needs of patients from different cultures. Cultural competence not only enhances patient satisfaction but also improves the quality of care, reduces health disparities, and promotes better health outcomes. In this section, we will explore the importance of cultural competence in healthcare delivery and discuss strategies to promote it.

Understanding Cultural Competence

Cultural competence is rooted in the recognition that cultural factors significantly influence health and wellness. It goes beyond mere cultural sensitivity or awareness and requires healthcare providers to actively engage in self-reflection, education, and the ongoing development of cultural knowledge and skills.

To provide culturally competent care, healthcare providers need to understand the cultural values, norms, and beliefs that shape patients' experiences and health-seeking behaviors. This includes considering factors such as language, religion, spirituality, socioeconomic status, gender, sexual orientation, and cultural practices.

Principles of Cultural Competence

Cultural competence is guided by several key principles that inform healthcare providers' interactions with patients. These principles include:

1. Respect for diversity: Acknowledging and valuing the diversity of patients' cultural backgrounds, experiences, and perspectives.

2. Self-awareness: Recognizing one's own cultural biases, beliefs, and values and the potential impact they may have on patient care.

3. Cross-cultural communication: Developing effective communication skills to bridge cultural and linguistic barriers, including the use of interpreters when necessary.

4. Adaptation of care: Tailoring healthcare practices and interventions to align with patients' cultural preferences and needs.

5. Collaboration: Engaging in collaborative decision-making with patients and their families, involving them in the healthcare planning process.

These principles serve as a foundation for promoting cultural competency and should be integrated into all aspects of healthcare delivery.

Strategies for Promoting Cultural Competence

Promoting cultural competence requires a multifaceted approach that involves both individual healthcare providers and healthcare organizations. Here are some strategies that can be employed:

1. Education and training: Healthcare providers should receive comprehensive education and training in cultural competence, including cultural humility, cultural awareness, and cultural knowledge specific to the populations they serve. This can be achieved through workshops, professional development programs, and diversity and inclusion training.

2. Culturally diverse workforce: Healthcare organizations should strive to have a diverse workforce that reflects the communities they serve. This diversity fosters cultural understanding, strengthens cross-cultural communication, and promotes rapport between healthcare providers and patients.

3. Use of interpreters and cultural brokers: Language and cultural barriers can impede effective communication and understanding between healthcare providers and patients. Healthcare organizations should provide access to trained interpreters and cultural brokers who can facilitate communication and ensure accurate information exchange.

4. Culturally responsive care environments: Healthcare organizations should create welcoming and inclusive care environments that respect and incorporate patients' cultural values and practices. This can include displaying culturally diverse artwork, providing educational materials in multiple languages, and ensuring that facilities are accessible to diverse populations.

5. Community engagement: Collaborating with community organizations and leaders is essential to understanding the specific needs and healthcare challenges of different cultural groups. By engaging the community, healthcare providers can develop culturally tailored interventions and programs that address these needs effectively.

Addressing Challenges and Limitations

Promoting cultural competence in healthcare delivery comes with challenges and limitations that need to be acknowledged and addressed. Some of these challenges include:

+ Language barriers: Language differences can hinder effective communication and understanding. Healthcare organizations should prioritize language access services, such as interpreter services, translated materials, and multilingual staff.

+ Time constraints: The additional time required for cultural assessments and understanding can pose challenges in busy healthcare settings. Healthcare providers need to find a balance between efficient care delivery and addressing cultural factors.

+ Implicit biases: Cultural competence training should address implicit biases that may influence healthcare providers' perceptions and decision-making. Education and self-reflection can help healthcare providers become aware of their biases and work towards minimizing their impact on patient care.

Despite these challenges, the importance of cultural competence in healthcare delivery cannot be overstated. By understanding and respecting the diverse cultural backgrounds of patients, healthcare providers can ensure equitable, patient-centered care that promotes better health outcomes.

Case Study: Cultural Competence in Maternal Health

Let's consider a case study to illustrate the importance of cultural competence in healthcare delivery. In many cultures, childbirth is considered a family event, with various rituals and practices surrounding it. However, in some healthcare settings, these cultural practices may be misunderstood or disregarded, leading to patient dissatisfaction and suboptimal care.

For example, in a culturally diverse community, a healthcare provider is caring for a pregnant woman who belongs to a culture where it is customary for extended family members to be present during labor and birth. However, due to hospital restrictions, only one support person is allowed in the delivery room. The healthcare provider, aware of the importance of cultural competence, engages in collaborative communication with the patient, her family, and the healthcare team to find a solution that respects the patient's cultural practices while addressing safety concerns. They explore alternatives such as allowing multiple support people during specific stages of labor or facilitating virtual participation through video calls. By understanding and accommodating the patient's cultural needs, the healthcare provider promotes a positive childbirth experience while ensuring high-quality care.

Conclusion

Cultural competence is a crucial component of healthcare delivery that promotes equitable and patient-centered care. By acknowledging and understanding the cultural diversity of patients, healthcare providers can eliminate barriers, reduce health disparities, and improve health outcomes. Implementing strategies to promote cultural competence, such as education, cultural diversity in the workforce, and collaboration with communities, can help healthcare organizations create inclusive environments that foster better patient-provider relationships and enhance the overall quality of care.

Research Ethics in Healthcare

Informed Consent and Research Participants' Rights

In the field of healthcare research, it is crucial to ensure that individuals who participate in research studies are fully informed about the nature and purpose of the study, as well as the potential risks and benefits involved. This process is known as informed consent. In this section, we will explore the concept of informed consent and discuss the rights of research participants.

Understanding Informed Consent

Informed consent is a fundamental ethical principle that guides the conduct of research involving human subjects. It is a process that involves providing individuals with the necessary information to make an informed decision about participating in a research study. The goal is to ensure that participants have a clear understanding of what the study entails, including the purpose, procedures, risks, benefits, alternatives, and their rights.

The process of obtaining informed consent typically involves several key elements. First, the researcher should provide a clear and understandable explanation of the study, including its objectives and procedures. This should be done in a language and format that is accessible to the participant, taking into account their cultural background and literacy level.

Next, the researcher should disclose any potential risks or discomforts associated with participating in the study. This may include physical, psychological, social, or economic risks. It is essential to provide a realistic assessment of these risks and to ensure that participants have a clear understanding of the potential consequences.

Additionally, the researcher should discuss the potential benefits of the study, both for the individual participant and for society as a whole. This can include advancements in knowledge, improvements in healthcare practices, or the development of new treatments or interventions.

Furthermore, the researcher should explain any alternatives to participation in the study, including any existing standard treatments or other research studies that may be available. Participants should be given the opportunity to ask questions and seek clarification about any aspect of the study.

Finally, informed consent involves ensuring that participants are aware of their rights as research subjects. This includes the right to withdraw from the study at any time without penalty or repercussion. Participants should also be informed about how their data will be handled and protected, and any privacy concerns should be addressed.

Ethical Considerations in Informed Consent

Obtaining informed consent is not simply a procedural requirement but a fundamental ethical obligation. It reflects the principles of autonomy, respect for persons, and the protection of individual rights. Without proper informed consent, research studies risk violating these principles and compromising the trust and wellbeing of research participants.

One critical aspect of informed consent is the voluntariness of participation. Participants must enter into the study willingly, without any coercion or undue influence. This means that researchers should avoid any form of manipulation or pressure that may sway a participant's decision-making process. In situations where vulnerable populations are involved, such as children, prisoners, or individuals with cognitive impairments, additional safeguards may be necessary to ensure their autonomy and protection.

Another ethical consideration is the comprehensibility and adequacy of the information provided during the informed consent process. The information should be presented in a way that is understandable to the participant, avoiding jargon or complex terminology. The researcher should also take into account any cultural or linguistic barriers that may impact the participant's understanding.

Furthermore, the ongoing nature of informed consent should be acknowledged. Participants may need additional information or have further questions even after providing initial consent. Researchers should be responsive to these needs and ensure that participants have access to continued support and clarification throughout the duration of the study.

Challenges and Solutions

Obtaining informed consent can present challenges in certain research contexts. For example, in clinical trials involving placebo controls, participants may need to be informed that they have a chance of receiving a placebo rather than the active treatment. This can pose challenges in terms of managing expectations and addressing concerns about receiving a potentially less effective treatment. To address this, researchers should carefully explain the rationale for including a placebo control, ensuring that participants understand the scientific and ethical justifications.

In research studies involving vulnerable populations, such as minors or individuals with cognitive impairments, obtaining informed consent can be particularly complex. In these cases, additional measures may be necessary to safeguard the rights and well-being of participants. This can include obtaining proxy consent from a legally authorized representative or using additional methods, such as simplified language or visual aids, to facilitate comprehension.

Collaborative approaches, such as community-based participatory research, can also enhance the informed consent process. By involving community members in the research design and implementation, researchers can ensure that the study aligns with the values and needs of the target population. This can contribute to a more meaningful and respectful informed consent process.

Real-World Examples

To illustrate the importance of informed consent, let's consider a real-world example. Imagine a clinical trial testing a new medication for a particular condition. During the informed consent process, participants should be provided with information about the potential side effects of the medication, such as nausea, headache, or dizziness. They should also be made aware of any potential risks associated with the study, such as the need for blood tests or additional doctor visits.

In another scenario, let's consider a genetic research study involving a population affected by a rare genetic disorder. In this case, researchers should provide detailed information about the study objectives, the process of genetic testing, and the potential implications of the results. Participants should be given the opportunity to ask questions and seek genetic counseling if needed.

Resources and Exercises

If you are interested in learning more about informed consent and research participants' rights, the following resources may be helpful:

- The Belmont Report: Ethical Principles and Guidelines for the Protection of Human Subjects of Research

- The Nuremberg Code

- The Declaration of Helsinki

- National Institutes of Health (NIH) resources on informed consent

- Institutional Review Board (IRB) guidelines and regulations

To further explore the topic, consider the following exercises:

1. Select a research study from a reputable journal and analyze the informed consent process. Did the study provide sufficient information about the purpose, procedures, risks, and benefits? Were the rights of the research participants adequately protected? Identify areas for improvement.

2. Imagine you are the principal investigator of a research study. Develop a comprehensive informed consent form that includes all the necessary information while ensuring it is understandable to a diverse group of potential participants.

3. Explore the ethical considerations in obtaining informed consent in a specific research context, such as genetic research, clinical trials, or social-behavioral studies. Discuss the challenges that may arise and potential strategies for addressing them.

Remember, the informed consent process is central to the ethical conduct of research and the protection of human rights. By ensuring individuals have the necessary information to make informed decisions, researchers can uphold the principles of autonomy, respect, and beneficence in their work.

Ethical Considerations in Clinical Trials

Clinical trials are essential for the development of new drugs, treatments, and medical interventions. They offer the opportunity to test the safety and efficacy of these interventions in human subjects before they are made available to the general population. However, conducting clinical trials also raises a number of ethical considerations that must be carefully addressed to ensure the protection of participants and the integrity of the research process. In this section, we will explore some of the key ethical issues that arise in the context of clinical trials.

Informed Consent and Research Participants' Rights

One of the fundamental ethical principles in clinical trials is the requirement for informed consent. Informed consent ensures that individuals have a thorough understanding of the study and its potential risks and benefits before deciding whether to participate. It is the responsibility of the researchers to provide clear and comprehensible information about the purpose of the trial, the procedures involved, the potential risks and benefits, and any alternative treatment options.

The informed consent process should be conducted in a culturally sensitive manner, taking into account the educational level and language proficiency of the participants. It should also include an opportunity for participants to ask questions and seek clarification. Informed consent should be voluntary, without any coercion or undue influence, and participants should have the freedom to withdraw from the trial at any time without penalty.

Research participants also have the right to privacy and confidentiality. Personal information collected during the trial should be kept confidential and should only be disclosed with the participant's permission or as required by law. Proper precautions should be taken to safeguard the privacy and confidentiality of participants' data.

Ethical Considerations in Clinical Trial Design

The design of a clinical trial plays a crucial role in ensuring the validity and ethical conduct of the research. Several ethical considerations come into play during the design phase.

Firstly, the selection of participants should be fair and unbiased. Researchers must avoid any form of discrimination or exclusion based on factors such as race, gender, age, or socioeconomic status. The inclusion and exclusion criteria should be justified by the scientific objectives of the trial and should not disproportionately exclude any particular group.

Secondly, the risks and benefits of the trial should be carefully evaluated. Researchers should conduct a thorough risk assessment to determine the potential harm that participants may face. This information should be communicated clearly to participants during the informed consent process. The potential benefits of the trial should also be weighed against the risks to ensure a favorable risk-benefit ratio.

Additionally, the control group in a clinical trial raises ethical considerations. While the use of a control group is necessary to establish the effectiveness of the intervention being studied, it may raise concerns about withholding potentially beneficial treatment from participants. In such cases, researchers should carefully consider the ethical justifications for using a control group and should ensure that participants in the control group receive appropriate and standard care.

Ethics in Recruitment and Incentives

Recruitment of participants is a critical aspect of clinical trials. It is essential to recruit a diverse and representative sample to ensure the generalizability of study findings. However, recruitment practices should be conducted ethically and should not exploit vulnerable populations or exert undue influence.

Incentives are often used to encourage participation in clinical trials. While incentives can be an effective way to motivate potential participants, they should be reasonable and should not unduly influence individuals to participate against their better judgment. Care should be taken to avoid coercive practices that may compromise the voluntary nature of informed consent.

Monitoring and Oversight

Ethical considerations extend beyond the initial stages of a clinical trial and continue throughout the study. Proper monitoring and oversight mechanisms are crucial to ensure the ongoing safety and welfare of participants.

An independent ethics committee or institutional review board (IRB) should review and approve the trial protocol to assess its scientific and ethical validity. The IRB should also conduct regular reviews and inspections to ensure that the trial is being conducted according to the approved protocol and in compliance with ethical guidelines.

Furthermore, proper monitoring of the trial should be conducted by qualified individuals who can promptly identify any adverse events or safety concerns. Regular evaluation of trial data should be conducted to ensure the ongoing scientific validity and ethical conduct of the study.

Addressing Conflict of Interest

Conflict of interest is another important ethical consideration in clinical trials. Researchers, sponsors, and institutions involved in the trial may have financial or other competing interests that could potentially compromise the integrity of the research.

Full disclosure of any conflicts of interest is essential to maintain transparency and trust. Researchers should declare any financial or non-financial relationships that could influence their objectivity or decision-making. This includes disclosing any financial support or affiliations with pharmaceutical companies or other interested parties.

Institutional policies and regulatory bodies should have mechanisms in place to identify, manage, and mitigate conflicts of interest. This may include requiring researchers to recuse themselves from decision-making processes or taking steps to ensure independent oversight of the trial.

Key Principles

When considering the ethical aspects of clinical trials, several key principles emerge:

+ **Respect for autonomy:** Participants' right to make informed decisions about their participation in a clinical trial should be respected.

+ **Beneficence and non-maleficence:** The potential benefits of the trial should outweigh the risks, and participants should not be exposed to unnecessary harm.

+ **Justice:** Fairness, equity, and non-discrimination in the selection and treatment of participants.

+ **Privacy and confidentiality:** Protection of participants' personal information and ensuring their privacy rights are respected.

+ **Transparency and accountability:** Ensuring that the trial is conducted in a transparent manner, with clear reporting of results, and accountability for researchers and sponsors.

Case Study: The Tuskegee Syphilis Study

A poignant example of unethical clinical research is the Tuskegee Syphilis Study conducted in the United States from 1932 to 1972. The study involved withholding treatment for syphilis from African American men, even after the discovery of effective treatment with penicillin. The participants were not adequately informed about the nature of the study, and their autonomy and dignity were violated.

The Tuskegee study serves as a powerful reminder of the importance of ethical principles in research and the need to safeguard the rights and welfare of research participants. It contributed to the establishment of ethical guidelines and regulations to prevent similar abuses from occurring in the future.

Conclusion

Ethical considerations in clinical trials are essential to protect the rights and welfare of research participants and to ensure the integrity and validity of the research process. Informed consent, fair participant selection, appropriate trial design, monitoring and oversight, and addressing conflicts of interest are key elements in ethical clinical trial conduct. By adhering to these principles, researchers and institutions can uphold the ethical standards necessary for the advancement of science and the improvement of health and wellness.

Conflict of Interest in Healthcare Research

In healthcare research, conflicts of interest occur when there is a divergence between the personal interests of individuals involved in the research and their professional responsibilities. These conflicts can potentially compromise the integrity and objectivity of the research, leading to biased or misleading results. It is crucial to identify and manage conflicts of interest to uphold the ethical standards and credibility of healthcare research.

Understanding Conflict of Interest

Conflict of interest arises when researchers, sponsors, or other stakeholders have personal, financial, or professional interests that may unduly influence their decision-making or research outcomes. These interests may include financial investments, employment relationships, consulting fees, ownership of intellectual property, or personal relationships. Conflicts of interest can occur at various stages of the research process, such as study design, data collection, analysis, and publication.

Conflicts of interest can be categorized into three broad types:

1. Financial Conflict of Interest: This type of conflict arises when researchers or institutions have financial ties, such as funding from pharmaceutical companies,

medical device manufacturers, or other commercial entities. Financial interests can lead to biased research results that favor the interests of the financial sponsor.

2. Intellectual Conflict of Interest: Intellectual conflicts of interest occur when researchers have personal beliefs, academic aspirations, or research commitments that may compromise their objectivity. These conflicts may arise from the desire to protect one's professional reputation or promote a particular hypothesis or theory.

3. Personal Conflict of Interest: Personal conflicts of interest involve non-financial motivations or relationships that may influence research integrity. These can include personal relationships with research participants, emotional attachments to certain outcomes, or political and ideological biases.

Impact of Conflict of Interest

Conflict of interest in healthcare research can have serious implications on both the research process and the interpretation of study findings. Some of the key impacts include:

1. Biased Study Design: Researchers with conflicts of interest may design studies that are more likely to produce results favorable to their interests, leading to skewed findings and inappropriate conclusions.

2. Selective Reporting: Conflicted researchers may selectively report or withhold data to present a more positive or desirable outcome. This can distort the overall evidence base and mislead decision-making in healthcare.

3. Influence on Peer Review and Publication: Researchers with conflicts of interest may try to influence the peer-review process or publication of research results, leading to biased dissemination of findings.

4. Undermining Public Trust: The presence of undisclosed conflicts of interest erodes public trust in the integrity and credibility of healthcare research. This can have long-term consequences for the translation of research into practice and policy-making.

Managing Conflict of Interest

To address conflicts of interest effectively, researchers, institutions, and funding agencies must adopt robust strategies for identification, disclosure, and management. Here are some key approaches:

1. Disclosure and Transparency: Researchers should disclose all potential conflicts of interest to ensure transparency. This includes financial relationships, competing professional commitments, and personal biases. Disclosures should be

made in research publications, presentations, and institutional review board (IRB) applications.

2. Independent Oversight: Establishing independent oversight bodies to review and manage conflicts of interest can help ensure objectivity. These bodies can assess and adjudicate potential conflicts, develop policies and guidelines, and monitor compliance.

3. Conflict Management Plans: Developing conflict management plans can help researchers and institutions proactively address conflicts of interest. These plans may include strategies such as recusing conflicted individuals from decision-making, establishing firewalls to separate conflicting interests, or providing alternative funding mechanisms to minimize financial conflicts.

4. Peer Review and Publication Practices: Journals and conference organizers should implement robust peer review processes that consider potential conflicts of interest. They should also have clear policies for editorial independence and disclosure of conflicts by authors, reviewers, and editors.

5. Ethical Training and Education: Researchers and healthcare professionals should receive comprehensive training on the identification, disclosure, and management of conflicts of interest. This can help promote a culture of ethical conduct and responsible research practices.

Case Study: Conflicts of Interest in Pharmaceutical Research

The influence of conflicts of interest is most evident in pharmaceutical research. One notable case is the opioid crisis in the United States, where conflicts of interest played a significant role. Pharmaceutical companies sponsored research studies promoting the widespread use of opioids for chronic pain management. These studies downplayed the risk of addiction and exaggerated the benefits of opioids, leading to the over-prescription of these drugs. Many of the researchers involved had financial ties to the pharmaceutical industry, which compromised the objectivity and reliability of their research.

This case underscores the importance of robust conflict of interest management in healthcare research. It also highlights the need for transparency and accountability in the pharmaceutical industry to prevent the undue influence of financial interests on research outcomes.

Conclusion

Conflict of interest is a complex ethical issue that requires careful attention in healthcare research. Transparency, disclosure, and independent oversight are essential for identifying and managing conflicts to preserve the integrity of research results. By addressing conflicts of interest effectively, we can maintain public trust, ensure the credibility of healthcare research, and promote evidence-based decision-making for the benefit of patients and society as a whole.

Data Privacy and Confidentiality in Research

In the field of health and wellness research, protecting the privacy and confidentiality of research participants is of paramount importance. Data privacy refers to the protection of personal information collected during research, while confidentiality refers to the obligation to keep that information secure and prevent unauthorized access. Researchers must adhere to ethical guidelines and legal requirements to ensure the privacy and confidentiality of research participants' data.

The Importance of Data Privacy and Confidentiality

Data privacy and confidentiality are crucial for several reasons. First and foremost, they respect the autonomy and dignity of research participants, as individuals have a right to control their personal information. Respecting privacy and maintaining confidentiality also help establish trust between researchers and participants, encouraging participation and minimizing the risk of social harm. Furthermore, ensuring data privacy is necessary to comply with legal and regulatory requirements, such as the General Data Protection Regulation (GDPR) and the Health Insurance Portability and Accountability Act (HIPAA).

Principles of Data Privacy and Confidentiality

Several principles guide the protection of data privacy and confidentiality in research:

1. **Informed Consent:** Research participants must provide informed consent before their data is collected, ensuring they understand the purpose of the research, the potential risks and benefits, and how their data will be used and protected. Informed consent forms should clearly outline the measures in place to protect data privacy and confidentiality.

2. **Anonymization and De-identification:** Researchers should take steps to remove any identifying information from research data to ensure participant anonymity. This can include using unique identifiers instead of personal identifiers and removing or encrypting sensitive information.

3. **Secure Data Storage:** Research data should be securely stored to prevent unauthorized access. This includes using encrypted storage devices, password-protected databases, and secure servers. Researchers must also implement appropriate security measures to protect against data breaches or cyber-attacks.

4. **Limited Data Access:** Researchers should limit access to research data to those individuals who require it for the purposes of the study. Access should be granted on a need-to-know basis, and data sharing agreements should outline the responsibilities of individuals with access to the data.

5. **Data Retention and Destruction:** Researchers should only retain research data for as long as necessary and should establish clear guidelines for data retention and destruction. When data is no longer required, it should be securely destroyed to prevent any potential breaches.

Challenges and Solutions

Protecting data privacy and confidentiality in research poses several challenges. As technology advances and data collection becomes more complex, the risk of data breaches increases. Additionally, ensuring the privacy and confidentiality of research participants becomes more challenging when conducting research with vulnerable populations or in global research collaborations.

To address these challenges, researchers can employ several strategies:

1. **Data Encryption:** Encrypting research data using robust encryption algorithms ensures that even if the data is accessed, it remains unreadable without the decryption key. Encryption should be used during data transmission and storage to prevent unauthorized access.

2. **Secure Data Sharing:** When collaborating with other researchers or sharing data with external parties, secure data sharing protocols should be established. This may involve using secure file-sharing platforms, ensuring data is encrypted during transit, and signing data-sharing agreements that outline the responsibilities and restrictions of data usage.

3. **Risk Assessment:** Conducting regular risk assessments can help identify potential vulnerabilities and implement appropriate safeguards. Researchers should assess the potential risks to data privacy and confidentiality throughout the research process and address these risks by implementing security measures and privacy-enhancing technologies.

4. **Training and Education:** Researchers and research staff should receive training on data privacy and confidentiality to ensure they understand their responsibilities and how to handle data securely. This includes training on secure data collection, storage, and transmission, as well as awareness of potential risks and best practices.

Moreover, the increasing use of Big Data and machine learning techniques in research raises additional challenges for data privacy and confidentiality. Researchers must be vigilant in ensuring that individual privacy is protected when analyzing large datasets. Techniques such as differential privacy, which adds noise to the data to protect individual privacy while still allowing for meaningful analysis, can be employed to address these challenges.

Case Study: Protecting Data Privacy in a Clinical Trial

Consider a case where researchers are conducting a clinical trial to evaluate the effectiveness of a new medication for a particular health condition. The trial involves collecting sensitive medical information from participants, including their medical history, test results, and treatment responses. In this case, protecting data privacy and confidentiality is crucial.

To ensure data privacy, the researchers implement strict protocols. They obtain informed consent from each participant, clearly explaining the purpose of the study, the potential risks and benefits, and the measures in place to protect their data. The researchers anonymize the data by assigning unique identifiers to participants instead of using personal identifiers, thus ensuring participant anonymity.

In terms of data storage, the researchers use encrypted storage devices and password-protected databases to securely store the data. They limit access to the data by designing the database with restricted user access rights and granting access only to authorized personnel involved in the study. Any data transmission is performed securely using encrypted channels.

The researchers also establish guidelines for data retention and destruction. They retain the data only for the duration of the study and securely destroy it once it is no longer needed, following proper data destruction protocols to prevent any potential breaches.

Throughout the study, the researchers conduct risk assessments to identify any vulnerabilities in data security. They stay updated with the latest guidelines and regulations regarding data privacy and confidentiality, and they provide training to study staff on secure data handling.

By implementing these measures, the researchers ensure that the privacy and confidentiality of research participants' data are protected, promoting trust, and complying with ethical and legal requirements.

Conclusion

Data privacy and confidentiality are paramount in health and wellness research. Researchers must adhere to ethical guidelines and legal requirements to protect the privacy of research participants and maintain the confidentiality of their data. Implementing measures such as informed consent, anonymization, secure data storage, limited data access, and proper data retention and destruction are essential to ensure data privacy and confidentiality. Additionally, addressing challenges such as technological advancements and vulnerabilities in data security requires regular risk assessments, training, and the use of privacy-enhancing technologies. By prioritizing data privacy and confidentiality, researchers can conduct research that is ethical, respectful, and trustworthy.

Ethical Issues in Reproductive Health and Genetic Technologies

Reproductive Rights and Autonomy

Abortion and Access to Reproductive Healthcare

Abortion is a complex and highly debated topic that touches upon moral, ethical, and legal considerations. It involves the termination of a pregnancy before the fetus reaches viability. Access to safe and legal abortion has been a crucial aspect of reproductive healthcare for women around the world. In this section, we will explore the ethical issues surrounding abortion, including the right to reproductive healthcare, the autonomy of pregnant individuals, and the balance between fetal and maternal interests.

The Right to Reproductive Healthcare

The right to reproductive healthcare is a fundamental aspect of individual autonomy and bodily integrity. It encompasses the right to make decisions about one's own reproductive health, including decisions about contraception, pregnancy, and abortion. This right is grounded in principles of human rights and social justice, recognizing that reproductive freedom is essential for gender equality and the overall well-being of individuals.

International human rights conventions, such as the International Covenant on Civil and Political Rights and the Convention on the Elimination of All Forms of Discrimination Against Women, affirm the right to access safe, legal, and affordable abortion services. However, the extent to which this right is protected and enforced varies among different countries and jurisdictions.

Autonomy and Decision-Making

Autonomy, or the right to self-determination, plays a central role in discussions about access to abortion. Pregnant individuals have the right to make decisions regarding their own bodies and reproductive lives. This includes the right to choose whether to continue or terminate a pregnancy.

The ethical principle of respect for autonomy supports the notion that individuals should have control over their own reproductive choices. Providing access to safe and legal abortion services allows pregnant individuals to exercise their autonomy and make decisions that align with their personal circumstances, values, and life plans.

Fetal Interests and Maternal Interests

The ethical debate around abortion also involves considering the interests of the fetus and the pregnant individual. Supporters of fetal rights argue that the fetus has moral status and should be afforded legal protection from harm. They contend that abortion is morally wrong because it terminates a potential life and infringes upon the rights of the fetus.

On the other hand, proponents of reproductive rights argue that the interests and well-being of the pregnant individual must be given equal consideration. They highlight the physical, emotional, and socioeconomic factors that can impact a person's ability to carry a pregnancy to term. They emphasize that a woman's right to her own body should not be superseded by the potential rights of the fetus.

Legal Frameworks and Access

The legal status of abortion varies widely across countries and jurisdictions. Some countries permit unrestricted access to abortion, while others impose restrictive laws that severely limit access. Legal frameworks may place gestational limits, require waiting periods, mandate counseling, or impose other barriers to accessing abortion.

Restrictive laws and limited access to abortion services can have substantial consequences. Pregnant individuals who are unable to access safe and legal abortions may resort to unsafe methods that endanger their health and lives. This highlights the importance of ensuring that reproductive healthcare services, including abortion, are available, accessible, and affordable to all individuals who need them.

Real-World Examples

To understand the complexity of abortion access, let's consider a couple of real-world examples:

1. In Country X, abortion is illegal except in cases where the woman's life is at risk. This strict legal framework limits access to safe abortions and forces many women to seek out clandestine and unsafe procedures. As a result, maternal mortality rates are disproportionately high in this country.

2. In Country Y, abortion is legal and accessible without restrictions up to a certain gestational age. However, barriers such as high costs, lack of information, and stigma prevent some individuals from accessing these services. This creates disparities in access and reinforces socioeconomic inequalities.

These examples illustrate the importance of not only legalizing abortion but also ensuring that it is accessible, affordable, and supported by comprehensive reproductive health services.

Resources and Support

Organizations like the World Health Organization (WHO), International Planned Parenthood Federation (IPPF), and Center for Reproductive Rights (CRR) work actively to promote and protect access to safe and legal abortion services worldwide. They provide resources, information, and support for individuals seeking reproductive healthcare, including abortion.

Additionally, healthcare providers and professionals play a crucial role in ensuring access to abortion services. They should be trained and supported to provide comprehensive, compassionate, and non-judgmental care to individuals seeking abortions, while also respecting their autonomy and confidentiality.

Exercise

Consider the following scenario:

Sarah, a 26-year-old woman, discovers that she is pregnant. She is in a stable relationship, financially secure, and has completed her education. However, she does not feel emotionally ready or prepared to become a parent at this stage in her life.

Identify and analyze the various ethical considerations and principles that come into play in this situation. Discuss the potential implications and consequences of different decisions, including reproductive healthcare options and access to abortion.

Conclusion

The issue of abortion and access to reproductive healthcare is nuanced and multifaceted. It is influenced by various ethical, legal, and social considerations. Protecting the right to access safe and legal abortion services is essential for promoting gender equality, individual autonomy, and reproductive freedom.

Ensuring access to comprehensive reproductive healthcare, including contraception, prenatal care, and abortion services, is vital for promoting the health and well-being of individuals and societies as a whole. It requires addressing the barriers and challenges that limit access, advocating for legal protections, and providing support and resources to those in need.

Assisted Reproductive Technologies and Parental Rights

Assisted reproductive technologies (ART) have revolutionized the field of reproductive medicine, offering hope to couples and individuals struggling with infertility. These technologies encompass a range of medical interventions that assist in the creation of a pregnancy, including in vitro fertilization (IVF), intracytoplasmic sperm injection (ICSI), and embryo cryopreservation. While ART provides new opportunities for individuals to become parents, it also raises important ethical considerations regarding parental rights and responsibilities.

The Ethics of Assisted Reproduction

The ethics of assisted reproduction revolve around issues such as access, safety, autonomy, and the well-being of both the parent(s) and the child. The underlying ethical principle is the promotion of the best interests of the child while respecting the rights and autonomy of the parents.

Autonomy and Informed Consent   A crucial ethical consideration in assisted reproduction is the principle of autonomy. Individuals and couples should have the right to make informed decisions about their reproductive choices, including whether to pursue ART and the specific procedures they want to undergo. Informed consent, which involves providing comprehensive information about risks, benefits, and alternatives, is essential in ensuring that individuals can make autonomous decisions. However, it is important to recognize that the complexity of reproductive technologies may pose challenges to fully understanding the risks and potential long-term consequences.

Beneficence and Non-Maleficence The principles of beneficence and non-maleficence guide healthcare providers and policymakers in promoting the well-being of individuals involved in assisted reproduction. The welfare of both the parent(s) and the future child must be carefully considered. Providers must ensure that medical procedures are performed safely and effectively, minimizing potential harm to all parties involved.

Parental Rights and Responsibilities Assisted reproductive technologies raise unique questions regarding parental rights and responsibilities. Traditional notions of biological and genetic ties are challenged by the use of donor gametes, surrogacy, and embryo adoption. Legal frameworks vary across jurisdictions, but commonly address issues of legal parental status, rights, and obligations. For example, in some countries, the gestational carrier may not have legal parental rights, while in others, a sperm or egg donor may retain certain rights or obligations.

Challenges and Ethical Dilemmas

The use of assisted reproductive technologies introduces complex ethical dilemmas that require careful consideration. These dilemmas often involve balancing the desires and rights of the parents with the best interests of the child.

Identity and Genetic Origins The use of donor gametes or embryos may raise questions about the child's genetic identity and their right to knowledge about their genetic origins. Disclosing or withholding this information can have important psychological and social implications for the child.

Surrogacy and Exploitation Surrogacy arrangements, where a woman carries a pregnancy for someone else, can present ethical concerns related to exploitation and the commodification of women's bodies. Questions arise regarding the autonomy and well-being of the surrogate, as well as the potential for undue influence in the establishment of the surrogacy agreement.

Access and Equity The cost and availability of assisted reproductive technologies can create disparities in access and equity. High costs may limit access to those with greater financial resources, raising questions of fairness and justice. In addition, the availability of these technologies varies across countries and regions, further exacerbating existing inequalities.

Ethical Frameworks for Assisted Reproductive Technologies

Several ethical frameworks can assist in navigating the complex issues surrounding assisted reproductive technologies.

Principle-Based Approach   Adopting a principle-based approach, such as the four principles of bioethics (autonomy, beneficence, non-maleficence, and justice), can provide a useful framework for analyzing the ethical considerations associated with assisted reproduction. This approach emphasizes balancing the competing interests and values of the various stakeholders involved.

Feminist Ethics   Feminist ethics critically examine the gendered dimensions of assisted reproductive technologies and consider the social, economic, and political factors that influence access and decision-making. This approach emphasizes the importance of reproductive autonomy and agency for all individuals involved, recognizing that different individuals may have different preferences and values regarding parenthood.

Case Study: Donor Conception

Consider the case of a couple who are unable to conceive a child naturally due to male-factor infertility. They opt for donor insemination to achieve pregnancy. The use of donor gametes raises questions about the child's right to genetic knowledge and the potential psychological impact of not having access to that information.

To address these concerns, healthcare providers should engage in a comprehensive informed consent process, ensuring that the couple fully understands the implications of using donor gametes. They should also offer counseling services to help the couple navigate the emotional and psychological complexities associated with donor conception. Additionally, legal frameworks should ensure mechanisms for the child to access genetic information, should they choose to do so later in life.

Resources and Further Reading

1. Baldwin, K., & Culley, L. (Eds.). (2019). Reproductive Donation: Practice, Policy, and Bioethics. Cambridge University Press.

2. Deonandan, R. (Ed.). (2018). Bioethics in Canada: A Philosophical Introduction. Canadian Scholars' Press Inc.

3. Glazer, N., & Baker, J. L. (Eds.). (2017). Access to Assisted Reproductive Technologies. Springer.

4. Smajdor, A., & Collier, R. (Eds.). (2017). An Introduction to Bioethics. John Wiley & Sons.

5. Human Fertilisation and Embryology Authority (HFEA). (n.d.). Guidance,   Policy,   and   Legislation.   Retrieved   from https://www.hfea.gov.uk/guidance/.

Ethical Considerations in Surrogacy

Surrogacy is a complex and highly debated topic in the field of reproductive health. It involves a woman carrying a pregnancy and giving birth to a child for another person or couple who cannot conceive or carry a child on their own. While surrogacy can be a wonderful option for individuals or couples seeking to become parents, it raises several ethical considerations that need careful examination. In this section, we will explore the ethical issues surrounding surrogacy and discuss various perspectives on the matter.

Defining Surrogacy

Before delving into the ethical considerations, it is important to define what surrogacy entails. There are two main types of surrogacy: gestational surrogacy and traditional surrogacy.

Gestational surrogacy involves the implantation of an embryo created using the egg and sperm of the intended parent(s) or donors into the uterus of the surrogate. In this case, the surrogate has no genetic connection to the child she carries.

On the other hand, traditional surrogacy involves the use of the surrogate's own egg, making her the biological mother of the child. This method is less common and raises additional ethical complexities compared to gestational surrogacy.

Both types of surrogacy present unique ethical considerations, and it is crucial to navigate them with sensitivity and respect for the well-being of all parties involved.

Autonomy and Informed Consent

One of the key ethical considerations in surrogacy is the concept of autonomy and informed consent. Surrogacy involves a complex decision-making process that requires full understanding and consent from all involved parties.

The intended parents must have a comprehensive understanding of the physical, emotional, and legal implications of the surrogacy arrangement. They should be fully informed about the potential risks, benefits, and uncertainties associated with the process. This includes the potential psychological and emotional impact on the surrogate, as well as the potential for complications during pregnancy and childbirth.

Equally crucial is ensuring that the surrogate fully understands the physical and emotional demands of surrogacy. She should provide informed consent and have the freedom to make decisions without any form of coercion or manipulation. It is essential that surrogates have access to independent legal counsel or medical professionals to guide them through the decision-making process.

Exploitation and Commodification

One of the main ethical concerns surrounding surrogacy is the potential for exploitation and commodification of women's bodies. Critics argue that the practice of paying women to carry a child for others can lead to an unequal power dynamic and the commodification of reproductive capabilities.

The commercialization of surrogacy has raised concerns about potential exploitation, especially in situations where surrogates may face financial or social pressures to enter into such arrangements. The need to protect vulnerable individuals from being taken advantage of is a significant ethical consideration in surrogacy.

To address these concerns, some jurisdictions have implemented regulations to protect the rights and well-being of surrogates. These regulations may include limiting compensation, implementing psychological screenings, and requiring legal contracts to ensure all parties are protected. Striking a balance between allowing surrogacy and preventing exploitation is a challenging task that requires careful consideration and ongoing evaluation.

Legal and Cultural Considerations

The ethical aspects of surrogacy are highly influenced by legal and cultural contexts. The legality of surrogacy varies widely across different countries and even within different states or provinces. Some jurisdictions have outright bans on surrogacy, while others permit only altruistic surrogacy arrangements.

The legal framework surrounding surrogacy must address several ethical questions, including the rights and responsibilities of all parties involved, the enforceability of contracts, and the determination of legal parentage. Creating appropriate legislation that protects the rights and interests of all individuals involved while upholding ethical principles is a complex task that requires interdisciplinary collaboration.

Additionally, cultural norms and values play a significant role in shaping the ethical considerations surrounding surrogacy. Some cultures may view surrogacy as a morally acceptable practice, while others may consider it morally objectionable.

Understanding and respecting cultural diversity is vital when discussing the ethical implications of surrogacy.

Child's Best Interests and Identity

Another important ethical consideration in surrogacy is the well-being and best interests of the child. Surrogacy raises unique questions about the child's identity, sense of belonging, and kinship relationships.

Children born through surrogacy may have a complex family structure that involves genetic parents, intended parents, and the surrogate. Questions regarding the child's right to know their genetic background, the disclosure of surrogacy to the child, and support for the child's emotional well-being must all be carefully addressed.

Guidelines and regulations should be established to ensure the child's best interests are prioritized throughout the surrogacy process. This may include providing counseling services, facilitating open communication, and promoting ongoing relationships between the child and all parties involved.

Conclusion

Surrogacy presents a multitude of ethical considerations that must be carefully evaluated to protect the well-being and rights of all individuals involved. Balancing the principles of autonomy, informed consent, prevention of exploitation, and the best interests of the child is crucial in navigating the ethical complexities of surrogacy.

A comprehensive approach that involves legal, medical, social, and ethical perspectives is necessary to develop regulations and guidelines that promote fairness, justice, and dignity for all parties involved. Ongoing research and dialogue in the field of surrogacy ethics are essential to ensure continual improvement in addressing these complex issues and promoting ethical and just practices in the realm of reproductive health.

Genetic Testing and Screening

Ethical Issues in Prenatal Genetic Testing

Prenatal genetic testing refers to medical procedures carried out during pregnancy to determine if a fetus has certain abnormalities or genetic conditions. These tests can provide valuable information to expectant parents about the health of their unborn child, but they also raise a number of ethical concerns. In this section, we will explore the ethical issues surrounding prenatal genetic testing and discuss the principles and considerations that guide decision-making in this area.

The Purpose and Types of Prenatal Genetic Testing

Before delving into the ethical issues, it is important to understand the purpose and types of prenatal genetic testing. Prenatal genetic testing aims to identify genetic disorders, chromosomal abnormalities, or other conditions that may affect the development or health of the fetus. There are two main categories of prenatal genetic tests: screening tests and diagnostic tests.

Screening tests, such as maternal serum screening and non-invasive prenatal testing, assess the chance of having a fetus with certain conditions. These tests can indicate an increased risk but cannot provide a definitive diagnosis. Diagnostic tests, such as chorionic villus sampling and amniocentesis, provide a more accurate diagnosis by analyzing fetal cells or samples. These tests carry a higher risk of complications compared to screening tests.

Now, let's explore some of the ethical issues that arise in the context of prenatal genetic testing.

Autonomy and Informed Decision-Making

One of the fundamental ethical principles in healthcare is autonomy, which refers to an individual's right to make informed decisions about their own healthcare. In the case of prenatal genetic testing, autonomy extends to the expectant parents who have the right to make decisions regarding the testing and the potential actions that may follow from the results.

However, informed decision-making in prenatal genetic testing can be complex. Expectant parents may feel overwhelmed by the information provided and the implications of different test results. It is critical that healthcare providers ensure that expectant parents have access to accurate and unbiased information about the purpose, benefits, limitations, and potential consequences of prenatal genetic testing. This includes information about the genetic conditions being tested for, the likelihood of false positives or false negatives, and the available options for further testing, treatment, or support.

Non-Directive Counseling

Non-directive counseling is an ethical principle that guides the communication between healthcare providers and patients in the context of prenatal genetic testing.

It emphasizes the importance of providing information and support without coercing or influencing the decision-making process of the expectant parents. Healthcare providers should strive to present information in a neutral and unbiased manner, ensuring that expectant parents understand the potential benefits, risks, and limitations of prenatal genetic testing. This allows the expectant parents to make their own decisions based on their personal values, beliefs, and circumstances.

Confidentiality and Privacy

Confidentiality and privacy play a crucial role in prenatal genetic testing. The results of these tests may have significant implications for the future of the unborn child and the expectant parents. It is essential that healthcare providers maintain the confidentiality of the test results, ensuring that they are only disclosed to the individuals who have a legitimate need to know.

Expectant parents should be informed about the confidentiality policies and procedures in place for prenatal genetic testing. They should also be given the opportunity to discuss any concerns or questions they may have about the privacy of their genetic information.

Psychological and Emotional Impact

Prenatal genetic testing can have a significant psychological and emotional impact on expectant parents. The potential for receiving a positive result indicating a genetic condition or abnormality can lead to increased stress, anxiety, and emotional distress.

Healthcare providers should be aware of the potential emotional impact of prenatal genetic testing and offer appropriate support and counseling services. It is crucial to address any concerns or questions the expectant parents may have and provide them with resources and referrals to support groups or mental health professionals, if necessary.

Selective Termination and Reproductive Decision-Making

One of the most ethically complex issues related to prenatal genetic testing is the possibility of selective termination, also known as selective abortion or selective feticide. Selective termination refers to the decision to end a pregnancy following the diagnosis of a serious genetic condition or abnormality.

Selective termination raises a host of ethical questions and dilemmas. Some individuals argue that it is a woman's right to make decisions about her own body and future, while others are concerned about the potential for discrimination against individuals with disabilities or the ethical implications of choosing which lives are worthy of being brought into the world.

Healthcare providers must approach discussions about selective termination with sensitivity and respect. They should provide comprehensive information about the available options, including continuing the pregnancy with appropriate support, adoption, or pursuing termination if desired. It is important to engage in open and non-judgmental conversations, allowing expectant parents to navigate the complexities and make decisions that align with their values and beliefs.

Resources and Support

Facing the ethical issues surrounding prenatal genetic testing can be challenging for expectant parents. It is crucial for healthcare providers to provide resources and support to help them navigate these complex decisions.

This may include referrals to genetic counselors, who can provide specialized information and guidance, as well as support groups or community organizations that can offer emotional support and advice. Open and ongoing communication between healthcare providers and expectant parents is key to ensure that all their needs are addressed throughout the process.

In conclusion, prenatal genetic testing raises a range of ethical issues, from autonomy and informed decision-making to privacy and the ethical complexities of selective termination. Healthcare providers must prioritize non-directive counseling, provide comprehensive information, maintain confidentiality and privacy, and offer appropriate psychological and emotional support. By doing so, they can assist expectant parents in making informed decisions that align with their values and beliefs, while navigating the ethical complexities of prenatal genetic testing.

Here is an example of the content for the section "4.2.2 Genetic Testing for Adults and Children":

Genetic Testing for Adults and Children

Genetic testing is a powerful tool that has the potential to revolutionize healthcare by providing valuable information about an individual's genetic makeup. It can help identify genetic variants that may increase the risk of developing certain diseases or conditions, as well as provide insights into an individual's response to certain medications.

Background

Genetic testing involves analyzing a person's DNA to detect changes or mutations in specific genes, chromosomes, or proteins. This information can be used to diagnose or predict disease, guide treatment decisions, inform reproductive choices, and assess the risk of certain conditions in both adults and children.

Ethical Considerations

While genetic testing offers great promise, it also raises several ethical considerations that need to be carefully addressed. One key concern is the potential for genetic discrimination. If the results of genetic testing were to be used against individuals, it could lead to the denial of employment, health insurance, or other opportunities. To prevent such discrimination, laws have been enacted, such as the Genetic Information Nondiscrimination Act (GINA) in the United States, which prohibits the use of genetic information in employment and health insurance decisions.

Another ethical consideration is the potential for psychological harm that may arise from the results of genetic testing. Learning about predispositions to certain diseases or conditions can cause anxiety, stress, and even depression. Genetic counselors play a crucial role in helping individuals understand and cope with the implications of genetic testing results, providing support and guidance throughout the process.

Principles of Informed Consent

Informed consent is a fundamental ethical principle that ensures individuals are adequately informed about the purpose, risks, and benefits of genetic testing before they decide to undergo the procedure. Genetic counselors play a crucial role in explaining the testing process, the potential outcomes, and the limitations of the results. They also help individuals understand the implications of the test results for themselves and their families.

Types of Genetic Testing

There are several types of genetic testing that can be performed on adults and children:

1. **Diagnostic Testing:** This type of testing is used to identify the cause of a specific genetic condition or disease. It is often performed when there is a

known family history of a genetic disorder or when an individual exhibits symptoms of a specific condition.

2. **Predictive Testing:** Predictive testing is used to determine an individual's risk of developing a genetic condition or disease later in life. This type of testing is often offered to individuals with a strong family history of a certain condition, such as certain types of cancer or neurological disorders.

3. **Carrier Testing:** Carrier testing is performed to identify individuals who carry a gene mutation for a genetic disorder, such as sickle cell anemia or cystic fibrosis. This type of testing is often recommended for individuals planning to start a family, as it can help assess the risk of passing on a genetic condition to their children.

4. **Pharmacogenetic Testing:** Pharmacogenetic testing aims to predict an individual's response to specific medications based on their genetic makeup. This information can help healthcare providers personalize medication prescriptions and avoid potential adverse reactions or inefficacy.

5. **Newborn Screening:** Newborn screening involves testing newborns for a variety of genetic and metabolic disorders shortly after birth. This screening allows for early detection and treatment of conditions that may not be apparent at birth but could have severe consequences if left untreated.

Case Study: Genetic Testing in Breast Cancer

To illustrate the impact of genetic testing, let's consider the case of genetic testing for breast cancer susceptibility genes, such as BRCA1 and BRCA2. Women who carry mutations in these genes have a significantly higher risk of developing breast and ovarian cancer.

Genetic testing for these mutations can help individuals make informed decisions about preventive measures, such as increased surveillance, prophylactic surgery, or chemoprevention. However, it also raises complex ethical issues, such as ensuring the privacy and confidentiality of test results, as well as the potential psychological impact of learning about one's increased risk for cancer.

In this case, genetic counselors play a crucial role in facilitating informed decision-making, providing emotional support, and guiding individuals through the available options. They help individuals understand their test results and discuss the implications for themselves and their families.

Conclusion

Genetic testing offers significant benefits in terms of disease prevention, personalized medicine, and reproductive choices. However, ethical considerations are essential to ensure that the technology is used responsibly and in the best interest of individuals and communities.

Genetic counselors play a vital role in ensuring informed consent, providing emotional support, and guiding individuals through the complex ethical considerations associated with genetic testing. As genetic technology continues to advance, it is crucial to have ongoing discussions and regulations in place to address emerging ethical challenges and ensure the ethical and responsible use of genetic testing in both adults and children.

Genetic Privacy and Discrimination

In this section, we will explore the ethical issues surrounding genetic privacy and discrimination. With advances in genetic technologies, individuals have greater access to information about their genetic makeup and potential health risks. However, this increased knowledge also raises concerns about privacy and the potential for discrimination based on genetic information.

The Importance of Genetic Privacy

Genetic privacy refers to the protection of an individual's genetic information from unauthorized access, use, or disclosure. This information includes genetic test results, family medical history, and other genetic data. Genetic privacy is essential for several reasons:

+ **Autonomy and Informed Decision-making:** Genetic privacy allows individuals to control who has access to their genetic information. It enables them to make informed decisions about sharing their information, including whether to undergo genetic testing or participate in research studies.

+ **Avoidance of Stigmatization:** Genetic information can reveal predispositions to certain diseases or conditions. Protecting genetic privacy helps prevent stigmatization or discrimination based on these vulnerabilities. Individuals may fear discrimination in employment, insurance coverage, or personal relationships if their genetic information is revealed.

+ **Protection of Family Members:** Genetic information not only affects the individual but also their family members who share genetic similarities. Maintaining genetic privacy ensures that individuals have control over the disclosure of information that may impact their relatives' lives.

Potential Forms of Genetic Discrimination

Genetic discrimination occurs when individuals are treated unfairly or unequally based on their genetic information. This discrimination can manifest in various ways:

+ **Employment Discrimination:** Employers may discriminate against individuals based on their genetic information, particularly if it suggests an increased risk of developing a certain condition. This discrimination can lead to hiring bias, denial of promotions, or termination.

+ **Insurance Discrimination:** Insurers may use genetic information to deny coverage or charge higher premiums to individuals who have an increased risk of developing certain conditions. This form of discrimination, known as genetic discrimination in insurance, can limit access to affordable health insurance.

+ **Education and Adoption Discrimination:** Genetic information can also be used to discriminate against individuals in educational settings or in adoption processes. It can lead to exclusion or denial of opportunities based on perceived genetic predispositions.

+ **Social and Personal Discrimination:** Genetic information, if disclosed, can lead to social stigma and personal discrimination. Individuals may face prejudice and exclusion from social groups or personal relationships due to their genetic makeup.

Legal and Ethical Considerations

To address concerns related to genetic privacy and discrimination, various legal and ethical frameworks have been established. These frameworks aim to protect individuals from harm and safeguard their rights.

+ **Genetic Information Nondiscrimination Act (GINA):** In the United States, the GINA prohibits health insurers and employers from discriminating against individuals based on their genetic information. It prevents the use of genetic information in employment decisions and restricts the use of genetic information in health insurance underwriting.

+ **Privacy Laws and Regulations:** Countries around the world have implemented privacy laws and regulations to protect genetic information. For example, the European Union's General Data Protection Regulation (GDPR) emphasizes the protection of genetic data as sensitive personal information.

+ **Informed Consent and Genetic Counseling:** Healthcare providers play a crucial role in ensuring genetic privacy and reducing discrimination risks. Genetic counseling before and after genetic testing helps individuals make informed decisions and understand the potential implications of genetic information disclosure.

Addressing Genetic Discrimination

To address genetic discrimination effectively, it is essential to take a multi-faceted approach. Here are some strategies that can help mitigate the risks of genetic discrimination:

+ **Education and Awareness:** Promoting public awareness about genetic privacy and discrimination can empower individuals to protect their rights and take informed actions. Education should focus on the importance of genetic privacy and the potential consequences of discrimination.

+ **Strengthening Legal Protections:** Continual evaluation and enhancement of existing legal protections, like GINA, can ensure they remain effective as genetic technologies advance. It is crucial to monitor and address any gaps in legislation to provide comprehensive coverage against genetic discrimination.

+ **Ethical Guidelines for Employers and Insurers:** Developing and implementing ethical guidelines for employers and insurers can help prevent genetic discrimination in these areas. These guidelines should emphasize fair decision-making and prohibit the use of genetic information in prejudiced ways.

+ **Data Security and Confidentiality:** Robust security measures should be in place to protect genetic data from unauthorized access. Healthcare institutions, researchers, and policymakers should prioritize data security and confidentiality to maintain public trust in the responsible use of genetic information.

+ **Advocacy and Support:** Organizations focused on genetic privacy and anti-discrimination advocacy can provide support for affected individuals and work towards enacting policies that prevent genetic discrimination. These organizations can provide resources, legal assistance, and platforms for individuals to share their experiences and raise awareness.

Case Study: GINA in Action

The Genetic Information Nondiscrimination Act (GINA) offers an example of legislation aimed at protecting individuals from genetic discrimination. One notable case involving GINA is the Burlington Northern Santa Fe Railway (BNSF) Company settlement in 2016. The Equal Employment Opportunity Commission (EEOC) filed a lawsuit against the BNSF after it required employees to undergo genetic testing and allegedly used the results to make employment decisions. The settlement resulted in the BNSF paying $2.9 million to affected employees and implementing new policies to prevent genetic discrimination.

This case highlights the importance of legal protections, like GINA, in safeguarding individuals' genetic privacy and preventing discriminatory practices. It also demonstrates the role of advocacy and legal action in holding organizations accountable for violating genetic privacy rights.

Exercises

1. Research the genetic privacy laws in your country or region. What specific provisions are in place to protect individuals from genetic discrimination?

2. Discuss the potential benefits and drawbacks of sharing genetic information in healthcare settings. How can healthcare providers balance the need for personalized treatment with the risks of genetic discrimination?

3. Investigate a real-life case where an individual experienced genetic discrimination. What were the consequences, and how could these have been prevented or addressed?

Resources

+ World Health Organization: *Human Genetics Program* - `https://www.who.int/genomics/public/geneticresearch/en/`

+ National Human Genome Research Institute: *Genetic Discrimination* - `https://www.genome.gov/10002077/genetic-discrimination-fact-sheet/`

+ American Society of Human Genetics: *Genetic Discrimination* - `https://www.ashg.org/policy/statements/13-ACMG-Discrimination.pdf`

In this section, we explored the importance of genetic privacy and the potential for discrimination based on genetic information. We discussed legal and ethical considerations and provided strategies to address genetic discrimination. The case study highlighted the impact of legislation like GINA in protecting individuals' genetic privacy. By understanding the ethical challenges and taking proactive measures, we can strive towards a future that respects genetic privacy and promotes equality in healthcare and other areas of life.

Human Enhancement Technologies

Ethical Considerations in Cosmetic Surgery

Cosmetic surgery, also known as aesthetic surgery, is a branch of medicine that focuses on enhancing a person's appearance through surgical procedures. It encompasses various treatments, including breast augmentation, liposuction, rhinoplasty (nose job), facelifts, and many others. While cosmetic surgery can bring positive changes to individuals' self-esteem and body image, it also raises important ethical considerations that need to be carefully addressed.

Balancing Autonomy and Societal Pressures

One of the key ethical considerations in cosmetic surgery is the balance between respecting patient autonomy and addressing the influence of societal pressures and unrealistic beauty standards. Patients have the right to make decisions about their bodies, including whether to pursue cosmetic procedures. However, it is crucial for healthcare professionals to engage in open and honest discussions with patients, ensuring that their motivations are well-founded and that they have realistic expectations.

Societal pressures and media influence can significantly impact patients' desires for cosmetic surgery. Unrealistic beauty standards portrayed in media can lead individuals to pursue surgery to conform to these standards rather than addressing their own self-esteem and body image issues. It is essential for healthcare professionals to be aware of these external factors and provide guidance to patients, emphasizing the importance of self-acceptance and mental well-being.

Informed Consent and Managing Expectations

Informed consent is a fundamental principle in any medical intervention, including cosmetic surgery. Patients must be adequately informed about the risks, benefits, alternative options, and expected outcomes of the procedure. It is the responsibility of healthcare professionals to ensure that patients have a comprehensive understanding of the potential risks, recovery process, and potential limitations of cosmetic surgery.

A crucial aspect of managing expectations is addressing the limitations of cosmetic surgery. While cosmetic procedures can enhance one's appearance, they cannot guarantee perfect results or fix underlying psychological issues. Healthcare professionals must educate patients on the importance of realistic expectations and the possibility of unforeseen complications or unsatisfactory results.

Minimizing Harms and Ensuring Patient Safety

Patient safety should always be the primary concern in any medical intervention, including cosmetic surgery. Healthcare professionals must prioritize the well-being and safety of patients and ensure that they are physically and emotionally fit for the procedure. Thorough medical evaluations, including assessments of mental health and emotional stability, should be carried out before proceeding with cosmetic surgery.

In addition to evaluating patients' medical suitability, healthcare professionals must minimize the risks associated with surgeries by adhering to strict standards of surgical practice. This includes maintaining a sterile surgical environment, utilizing appropriate anesthesia techniques, and closely monitoring patients during and after the procedure.

Addressing Socioeconomic Inequalities

Cosmetic surgery is often seen as a privilege accessible only to those with sufficient financial resources. This creates socioeconomic inequalities, as individuals from lower-income backgrounds may be unable to afford these procedures. Healthcare professionals should be mindful of these disparities and strive to promote equitable access to cosmetic surgery.

One approach to addressing socioeconomic inequalities is advocating for increased insurance coverage for medically necessary cosmetic procedures. Some individuals may require reconstructive surgeries due to congenital deformities, accidents, or medical conditions. Insurance coverage for such procedures can help bridge the gap and ensure that individuals with legitimate medical needs have access to the necessary treatments.

Furthermore, healthcare professionals should actively engage in education and public awareness campaigns to challenge societal beauty standards and promote self-acceptance. By fostering a culture that celebrates diverse body types and appearances, we can help reduce the demand for cosmetic surgery solely driven by societal pressures.

Ethical Considerations in Advertising and Marketing

Advertising and marketing practices in the cosmetic surgery industry raise ethical concerns. Healthcare professionals should adhere to ethical guidelines and avoid using misleading or overly persuasive advertising tactics that might exploit individuals' vulnerabilities or contribute to unrealistic beauty expectations.

Clear and accurate information should be provided regarding the risks, benefits, and limitations of cosmetic procedures in all advertising materials. Healthcare professionals should also ensure that informed consent is obtained before any procedure and that potential patients are not coerced or pressured into making hasty decisions.

Case Study: Social Media Influencers and Cosmetic Surgery

The rise of social media influencers and their impact on cosmetic surgery is a contemporary ethical dilemma. Many influencers promote cosmetic procedures, often without disclosing any conflicts of interest. This raises concerns about the authenticity of their recommendations and the potential influence on vulnerable individuals.

Healthcare professionals need to address this issue proactively. They should advocate for transparency and ethical practices in influencer marketing, encouraging influencers to disclose any financial relationships with cosmetic surgery providers. Additionally, healthcare professionals should provide evidence-based information to counteract the potential misinformation spread by influencers and advise patients to critically evaluate the credibility of the sources they follow.

Conclusion

Ethical considerations play a crucial role in the field of cosmetic surgery. Balancing patient autonomy with societal pressures, ensuring informed consent, prioritizing patient safety, addressing socioeconomic inequalities, and promoting ethical advertising practices are all necessary to navigate the complex landscape of cosmetic surgery. By integrating these principles into clinical practice, healthcare professionals can promote ethical decision-making and positive outcomes for patients seeking cosmetic interventions.

Cognitive Enhancement and Ethics

Advancements in science and technology have opened up new possibilities for enhancing cognitive abilities. Cognitive enhancement refers to the use of various interventions or techniques to improve cognitive functions such as memory, attention, and problem-solving. These enhancements can range from lifestyle modifications to the use of pharmaceutical drugs or brain stimulation techniques. However, the ethical considerations surrounding cognitive enhancement are complex and require careful examination.

Defining Cognitive Enhancement

Before delving into the ethical issues, let us first define what is meant by cognitive enhancement. Cognitive enhancement refers to any intervention aimed at improving cognitive function beyond what is considered normal or typical for an individual. This can include activities such as brain training exercises, the use of pharmacological substances, or even more advanced techniques such as neurofeedback or brain computer interfaces.

The Pursuit of Enhancement

The pursuit of cognitive enhancement raises several ethical questions. One of the main concerns is the potential for unequal access and distribution of enhancement technologies. If certain individuals or groups have better access to cognitive enhancement interventions, it could widen existing social and economic disparities. For example, if expensive pharmaceutical drugs are the primary mode of cognitive enhancement, only those who can afford them will have access to such enhancements, further exacerbating inequality.

Autonomy and Authenticity

Another ethical concern is the impact of cognitive enhancement on personal autonomy and authenticity. Autonomy refers to an individual's ability to make informed decisions and have control over their own lives. Some argue that by using cognitive enhancement technologies, individuals may be undermining their own autonomy as they rely on external interventions to achieve certain cognitive abilities.

Furthermore, the use of cognitive enhancement methods could raise questions about the authenticity of a person's achievements. If someone achieves exceptional cognitive abilities through enhancement techniques, can their achievements be considered genuinely theirs? This raises concerns about fairness in competitive settings such as education or employment, where individuals should be evaluated based on their innate talents and efforts rather than external cognitive enhancements.

Safety and Long-term Effects

The safety and long-term effects of cognitive enhancement interventions are also of ethical concern. While some enhancements, such as lifestyle modifications or brain training exercises, may have minimal risks, other interventions, such as pharmacological substances, may have unknown or adverse effects on individuals' health and well-being.

Additionally, the long-term effects of cognitive enhancements are not well understood. It is uncertain how these enhancements may affect individuals' cognitive abilities and overall functioning over an extended period. Ethical considerations should take into account the potential risks and uncertainties associated with cognitive enhancement interventions.

Equity and Access

The ethical principle of equity requires that individuals have equal opportunities and fair access to resources and opportunities. In the context of cognitive enhancement, it is crucial to ensure that enhancements are accessible to all individuals, regardless of their socio-economic status or background. This means that policies and regulations should be in place to prevent the unequal distribution and accessibility of cognitive enhancement technologies.

Furthermore, equity should also consider the potential impact of cognitive enhancements on marginalized or vulnerable populations. It is necessary to ensure that these technologies do not exacerbate existing social inequalities or contribute to further marginalization or stigmatization of certain groups.

Balancing Risks and Benefits

Ethical decision-making regarding cognitive enhancement should involve a careful evaluation of the risks and benefits associated with each intervention. It is essential to consider the potential benefits of improved cognitive abilities, such as enhanced learning or problem-solving skills, against the risks of potential adverse effects or unintended consequences.

Evaluating the risks and benefits should also involve considerations of individual well-being and societal implications. For example, if a particular cognitive enhancement intervention improves memory but compromises an individual's emotional well-being, the overall benefit may be questionable. Similarly, the societal impact of widespread cognitive enhancement should be carefully assessed to ensure that it aligns with broader values and goals.

Ethical Frameworks for Cognitive Enhancement

When considering the ethical issues related to cognitive enhancement, various ethical frameworks can provide guidance. Utilitarianism, for instance, focuses on maximizing overall well-being and may consider cognitive enhancements acceptable if they lead to net benefits for individuals and society.

On the other hand, deontological perspectives, such as Kantian ethics, prioritize respect for human dignity, autonomy, and the value of human beings as ends in themselves. From this perspective, there may be concerns about cognitive enhancement interventions that undermine individual autonomy or perpetuate social inequalities.

A virtue ethics approach may emphasize the development of virtues and character traits that lead to flourishing and well-being. Within this framework, cognitive enhancement interventions should be evaluated based on their impact on an individual's overall character and flourishing.

Case Study: Neuroenhancement Drugs

To illustrate some of the ethical considerations surrounding cognitive enhancement, let's examine the case of neuroenhancement drugs. These drugs, such as methylphenidate (commonly known as Ritalin) or modafinil, have been used off-label for cognitive enhancement purposes.

One ethical concern is the misuse or abuse of these drugs. Neuroenhancement drugs are prescription medications intended for medical conditions such as attention deficit hyperactivity disorder (ADHD) or narcolepsy. Using these drugs without a medical need may have adverse effects and raises concerns about ethical use.

Moreover, access to neuroenhancement drugs may further perpetuate existing social disparities. Individuals with financial resources and connections may have better access to these drugs, putting others at a disadvantage. This raises questions about fairness and equity in the use of cognitive enhancement interventions.

Additionally, the long-term effects of neuroenhancement drugs are not well understood. There may be risks and unknown consequences associated with the prolonged use of these substances. Ethical considerations should include a thorough evaluation of the potential risks and benefits of using neuroenhancement drugs.

Conclusion

Cognitive enhancement presents complex ethical challenges that need careful consideration. Balancing issues of fairness, autonomy, safety, and societal impact is crucial in evaluating the ethical implications of cognitive enhancement interventions.

To navigate these challenges, policymakers, ethicists, and researchers should engage in interdisciplinary discussions to develop guidelines and regulations that promote ethical and responsible cognitive enhancement practices. It is essential to ensure equitable access to cognitive enhancement technologies and prioritize individual autonomy and well-being while minimizing potential risks. By addressing these ethical concerns, we can foster a more just and ethically responsible approach to cognitive enhancement in the future.

Genetic Engineering and Ethical Implications

Genetic engineering is a field that involves modifying an organism's genetic material to alter its characteristics or introduce new traits. This technology has gained significant attention and raised numerous ethical concerns due to its potential to reshape the future of human health and wellness. In this section, we will explore the ethical implications associated with genetic engineering and discuss the key considerations that need to be addressed.

Understanding Genetic Engineering

Before delving into the ethical issues, it is essential to have a clear understanding of what genetic engineering entails. Genetic engineering involves manipulating an organism's DNA, either by adding, deleting, or modifying specific genes. This process allows scientists to introduce new traits, improve desired characteristics, or eradicate harmful traits in organisms.

Various techniques, such as gene editing tools like CRISPR-Cas9, have revolutionized genetic engineering and made it more accessible and precise. These techniques enable scientists to edit genes with unprecedented accuracy, leading to significant advancements in fields like medicine, agriculture, and environmental conservation.

Ethical Concerns

As with any powerful technology, genetic engineering raises numerous ethical concerns. These concerns stem from the potential consequences and implications of altering an organism's genetic makeup. Let's explore some of the key ethical implications associated with genetic engineering:

1. **Playing God**   One of the primary concerns raised about genetic engineering is the notion of humans "playing God." By manipulating the genetic material of

organisms, we are essentially taking control of nature and altering its course. This raises profound philosophical questions about our role and responsibility in shaping the natural world.

2. Unintended Consequences Genetic engineering has the potential to create unintended consequences. Modifying an organism's genetic makeup can have unpredictable effects on its health, behavior, and interactions with the environment. These unintended consequences could lead to unforeseen ecological disruptions or harm to human health.

3. Genetic Discrimination Genetic engineering opens the door to genetic discrimination. As we gain the ability to manipulate genetic traits, it raises concerns about discrimination based on an individual's genetic profile. This could manifest in various forms, such as denying employment or insurance coverage based on genetic predispositions to certain diseases.

4. Equity and Access The availability and affordability of genetic engineering technologies raise concerns about equity and access. If these technologies are only accessible to a privileged few, it could further exacerbate existing social and economic inequalities. There is a need to ensure fair and equitable distribution of genetic engineering advancements to avoid creating a genetic divide.

5. Informed Consent and Autonomy Genetic engineering can have implications for an individual's right to autonomy and informed consent. The ability to modify an individual's genetic makeup raises questions about who should have the authority to make decisions about genetic modifications and how much information should be disclosed to individuals undergoing genetic interventions.

6. Environmental Impact Genetic engineering can have far-reaching effects on the environment. Altering the genetic makeup of organisms can lead to unintended consequences, including the potential for genetically modified organisms (GMOs) to spread and disrupt natural ecosystems. The ecological impact of GMOs and the potential for irreversible environmental changes raise significant ethical concerns.

Addressing Ethical Concerns

To address the ethical concerns associated with genetic engineering, several key considerations need to be taken into account:

1. **Robust Ethical Frameworks** It is crucial to establish robust ethical frameworks and guidelines that govern the use of genetic engineering technologies. These frameworks should address concerns about responsible innovation, environmental impact, equitable access, and respect for individual autonomy.

2. **Public Engagement and Education** Engaging the public in discussions about genetic engineering can help foster informed decision-making and ensure that societal values and concerns are taken into account. Education about the science and ethics of genetic engineering is vital, as it allows individuals to critically evaluate its benefits and risks.

3. **Regulation and Oversight** Governments and international bodies play a crucial role in regulating and overseeing genetic engineering practices. Establishing clear regulations, guidelines, and monitoring mechanisms can ensure that genetic engineering is conducted responsibly, with proper risk assessment and consideration of ethical implications.

4. **International Cooperation** Genetic engineering is a global issue that requires international cooperation. Collaborative efforts can help address concerns related to global disparities in access to genetic engineering technologies, harmonize regulations, and promote ethical practices across borders.

Case Study: Genetically Modified Organisms

Genetically modified organisms (GMOs) serve as a notable example of the ethical implications of genetic engineering. GMOs have been extensively used in agriculture to enhance crop yield, pest resistance, and nutrient content. However, their use has also sparked concerns about environmental impact, food safety, and the consolidation of corporate control over the global food system.

The ethical considerations surrounding GMOs revolve around issues such as environmental impact, potential health risks, socioeconomic implications, and labeling transparency. Balancing the benefits and risks of GMOs requires careful assessment of their impact on ecosystems, consumer health, and the livelihoods of farmers.

Conclusion

Genetic engineering holds immense promise for improving human health and addressing pressing global challenges. However, the ethical implications must be carefully considered and addressed to ensure responsible and equitable use of these technologies. By establishing robust ethical frameworks, fostering public engagement, and promoting international cooperation, we can navigate the ethical complexities of genetic engineering and shape a more just and sustainable future for health and wellness.

Key Takeaways

- Genetic engineering involves modifying an organism's genetic material to alter its characteristics or introduce new traits. - Ethical concerns associated with genetic engineering include playing God, unintended consequences, genetic discrimination, equity and access, informed consent and autonomy, and environmental impact. - Addressing these concerns requires robust ethical frameworks, public engagement and education, regulation and oversight, and international cooperation. - GMOs serve as a case study highlighting the ethical implications of genetic engineering in agriculture. - Responsible and equitable use of genetic engineering technologies is crucial to shape a more just and sustainable future.

Bioethics and New Reproductive Technologies

In Vitro Fertilization and Ethical Dilemmas

In vitro fertilization (IVF) is a reproductive technology that involves the fertilization of an egg with sperm outside of the body. It is a widely used assisted reproductive technique that has revolutionized the field of reproductive medicine and has helped many couples and individuals who struggle with infertility to conceive a child. However, IVF is not without its ethical dilemmas and controversial issues.

Background

IVF was first successfully performed in the late 1970s and has since become a common treatment option for infertility. The procedure involves stimulating a woman's ovaries to produce multiple eggs, retrieving the eggs, fertilizing them in a laboratory dish, and then transferring the resulting embryos back into the woman's uterus. IVF can be used for various reasons, including tubal factor infertility, endometriosis, male factor infertility, and unexplained infertility.

Ethical Considerations

While the goal of IVF is to help individuals and couples achieve their desire for a child, there are several ethical considerations that arise in the context of this reproductive technology. These include concerns about the status of the embryo, the use of donor gametes, the potential for multiple pregnancies, and the moral and legal implications of embryo disposal.

Status of the Embryo   One of the main ethical dilemmas surrounding IVF is the moral status of the embryo. In IVF, multiple embryos are often created, but not all of them can be transferred into the woman's uterus. This raises questions about the moral status of the surplus embryos. Some argue that embryos have the same moral and legal status as fully-developed human beings, and therefore discarding or using them for research purposes is morally wrong. Others argue that embryos have a moral status that is lower than that of a fully-developed human being and that their use for reproductive purposes or research is ethically justifiable.

Use of Donor Gametes   IVF also raises ethical concerns regarding the use of donor gametes, including sperm and eggs. Donor gametes are often used when one or both partners have fertility issues, or in cases where a single individual desires to have a child. Ethical considerations arise regarding the anonymity of the donors, the rights and responsibilities of the donors and recipients, and potential implications for the child's identity and well-being.

Multiple Pregnancies   Another ethical dilemma associated with IVF is the risk of multiple pregnancies. To increase the chances of success, multiple embryos are often transferred into the woman's uterus. However, this can lead to a higher probability of multiple pregnancies, which pose health risks for both the mother and the fetuses. Multiple pregnancies are more likely to result in premature birth, low birth weight, and other complications. The ethical challenge lies in balancing the desire for a successful pregnancy with the potential harm to the mother and the unborn children.

Embryo Disposal   IVF procedures often result in the creation of more embryos than can be transferred or used for research purposes. This raises ethical questions about what should be done with the surplus embryos. Options include freezing them for future use, donating them to other individuals or couples, donating them for research purposes, or disposing of them. Each of these options has ethical considerations related to the value and dignity of the embryo, the rights and interests of the donors and recipients, and the potential implications for the individuals involved.

Addressing Ethical Dilemmas

Addressing the ethical dilemmas associated with IVF requires careful consideration of the values and principles involved. Ethical frameworks such as deontology, consequentialism, and virtue ethics can provide guidance for decision-making in this context.

Deontological Approach A deontological approach to the ethical dilemmas of IVF focuses on the inherent value and rights of the embryo. Advocates of this approach argue that embryos should be considered as persons from the moment of conception and should be afforded the same moral and legal protection as fully-developed human beings. From a deontological perspective, practices such as discarding surplus embryos or using them for research purposes may be considered morally wrong.

Consequentialist Approach A consequentialist approach considers the outcomes and consequences of IVF practices. Proponents of this approach argue that the overall well-being and happiness of the individuals involved should be the determining factor in decision-making. From a consequentialist perspective, the ethical acceptability of practices such as IVF, the use of donor gametes, or the disposal of surplus embryos depends on whether they result in positive outcomes and minimize harm.

Virtue Ethics Approach Virtue ethics emphasizes the development of virtuous character traits and the cultivation of moral virtues. In the context of IVF, a virtue ethics approach would consider the virtues of compassion, empathy, and justice. Practitioners and policymakers would need to consider how best to balance the desires and interests of the individuals involved, while also promoting the well-being and dignity of the embryos, donors, and recipients.

Case Study: Preimplantation Genetic Diagnosis

Preimplantation genetic diagnosis (PGD) is a technique used in IVF to screen embryos for genetic disorders before implantation. This raises ethical questions regarding the selection of embryos based on genetic traits and the potential for eugenics.

A couple, Mary and John, undergo IVF with PGD due to a family history of a severe genetic disorder. Through PGD, they are able to identify healthy embryos for transfer. However, they also discover that one of the embryos has a trait associated with a higher risk of developing a common health condition later in life, such as diabetes.

From a consequentialist perspective, Mary and John may be inclined to choose not to transfer the embryo with this trait, as they believe it would increase their future child's well-being and reduce the risk of developing a health condition. However, a deontological approach might argue that the embryo should be given the same moral consideration and respect as any other, regardless of genetic traits.

Virtue ethics would encourage Mary and John to consider the virtues of empathy and justice. They might reflect on the potential implications of their decision on the future child, as well as society's perception and treatment of individuals with the health condition in question.

Conclusion

In vitro fertilization is a complex reproductive technology that raises various ethical dilemmas. These include questions about the status of the embryo, the use of donor gametes, the potential for multiple pregnancies, and the moral and legal implications of embryo disposal. Addressing these dilemmas requires a careful consideration of ethical frameworks, such as deontology, consequentialism, and virtue ethics. Ultimately, the goal is to strike a balance between the desires and well-being of the individuals involved while promoting the principles of justice, autonomy, and respect.

Cloning and the Ethics of Reproductive Science

Cloning is a controversial topic that raises profound ethical questions about the nature of reproduction, the boundaries of science, and the value of human life. In this section, we will explore the ethical considerations surrounding cloning and reproductive science, including the different methods of cloning, the potential benefits and risks, and the social and moral implications.

Background

Cloning is the process of creating an organism that is genetically identical to another. There are three main types of cloning: gene cloning, reproductive cloning, and therapeutic cloning. In gene cloning, a specific gene or DNA sequence is copied and inserted into another organism. Reproductive cloning, on the other hand, involves creating a genetically identical copy of an existing organism. Lastly, therapeutic cloning aims to create embryos for the purpose of harvesting stem cells to treat diseases.

The most well-known method of reproductive cloning is somatic cell nuclear transfer (SCNT). This technique involves removing the nucleus of an egg cell and replacing it with the nucleus of a somatic cell, such as a skin cell, from the individual to be cloned. The resulting embryo is then implanted into a surrogate mother, where it develops into a clone of the original organism.

Principles and Ethical Considerations

The ethical considerations surrounding cloning and reproductive science revolve around several key principles:

1. **Human Dignity and Personhood:** Cloning raises questions about the intrinsic value and dignity of human life. Critics argue that creating clones diminishes the uniqueness and individuality of a person, reducing them to mere copies. Additionally, there are concerns about the potential for dehumanization and exploitation.

2. **Autonomy and Reproductive Liberties:** Supporters of cloning argue that individuals should have the right to make choices about their reproductive options, including the choice to clone themselves. They believe that reproductive liberties should extend to emerging technologies like cloning.

3. **Health and Safety:** Cloning has been associated with a range of health risks, such as developmental abnormalities, organ malfunction, and premature aging. Ethical considerations require evaluating the safety and well-being of potential clones and the risks associated with the cloning process.

4. **Social and Psychological Impact:** Cloning could have significant social and psychological implications. Cloned individuals might face unique challenges, including identity issues, familial relationships, and societal perceptions. Additionally, the availability of cloning technology could exacerbate social inequalities.

5. **Reproductive Justice:** Cloning raises questions about equitable access to reproductive technologies. The cost and availability of cloning procedures may create disparities, limiting access to those with financial means and exacerbating existing social inequalities.

Contemporary Issues and Debates

Although cloning has not yet been successfully achieved in humans, the prospect of human cloning has generated significant debate and speculation. Here are some of the key contemporary issues and debates surrounding cloning:

1. **Ethical Limits on Cloning Research:** Many countries have implemented legal frameworks to regulate cloning research. These frameworks aim to strike a balance between promoting scientific advancements and preventing unethical practices. The debates often center around the permissible uses of cloning technology, such as whether it should be limited to therapeutic purposes or completely prohibited.

2. **Cloning for Reproduction vs. Cloning for Research:** One of the major debates is whether cloning for reproduction should be allowed. The notion of creating genetically identical copies of individuals raises fears of a "cloning industry" or the potential for reproductive cloning to be used for nefarious purposes. On the other hand, cloning for research and therapeutic purposes, such as cultivating organs for transplantation, has garnered more support.

3. **Legal and Regulatory Frameworks:** Developing comprehensive legal and regulatory frameworks for cloning is crucial to address the ethical concerns and potential risks associated with the technology. These frameworks should consider issues such as informed consent, safety regulations, and the prevention of reproductive cloning.

4. **Public Perception and Acceptance:** Public opinion on cloning varies widely. Factors such as religious beliefs, cultural norms, and scientific literacy influence public perception. Ethical debates surrounding cloning should take into account public concerns and provide transparent and accessible information to foster informed discussions.

Case Study: The Cloning of Dolly the Sheep

An iconic case in the history of cloning is the cloning of Dolly the Sheep in 1996. Dolly, the first successfully cloned mammal from an adult somatic cell, ignited a worldwide debate on the ethics of cloning. The case raised concerns about animal welfare, the potential for human cloning, and the long-term health outcomes for clones.

Dolly's cloning demonstrated the scientific and technological capabilities of cloning but also sparked ethical discussions around the moral implications of creating clones, particularly sentient beings. The case study of Dolly the Sheep showcases the need for careful consideration of the ethical dimensions of cloning, including the potential consequences and risks involved.

Conclusion

The ethics of cloning and reproductive science are complex and multifaceted. The considerations surrounding the cloning process extend beyond mere scientific advancements and delve into the realms of human dignity, autonomy, justice, and societal impact. As technology continues to advance, it becomes imperative to navigate the ethical landscape of cloning, ensuring that scientific progress aligns with societal values and well-being. By critically evaluating the ethical concerns and engaging in open and inclusive dialogues, we can shape a more ethical and just future for reproductive science.

Gene Editing and the Future of Reproduction

In recent years, advancements in gene editing technology have opened up new possibilities in the field of reproduction. This technology, known as CRISPR-Cas9, allows scientists to make precise changes to the DNA of living organisms, including humans. This has raised important ethical considerations and has the potential to shape the future of reproduction.

The Science of Gene Editing

Gene editing involves modifying the DNA sequence of an organism's genome. CRISPR-Cas9, short for Clustered Regularly Interspaced Short Palindromic Repeats-CRISPR-associated protein 9, is a revolutionary gene editing tool that uses a guide RNA molecule to target specific locations in the genome and a protein called Cas9 to cut the DNA at those locations. Once the DNA is cut, the cell's natural repair mechanisms can be harnessed to introduce desired changes.

The possibilities of gene editing in reproduction are immense. It could potentially be used to correct genetic disorders, prevent the transmission of hereditary diseases, and enhance the genetic traits of future generations. However, these possibilities also raise significant ethical concerns that must be considered.

Ethical Considerations in Gene Editing

One of the primary ethical concerns surrounding gene editing in reproduction is the potential for "designer babies" or the deliberate manipulation of traits to create "perfect" offspring. This raises questions about the limits of parental autonomy and the potential for exacerbating existing social inequalities.

Another ethical consideration is the impact of gene editing on future generations. By making modifications to the germline cells (eggs and sperm), these changes would be heritable and could affect not only the individual being edited but also their descendants. This raises concerns about the potential unintended consequences and long-term effects of altering the human gene pool.

There is also a strong need to ensure the safety and efficacy of gene editing techniques. The technology is still relatively new, and further research is required to understand the potential risks and limitations. Additionally, ensuring equitable access to these technologies is crucial to prevent exacerbating existing disparities in healthcare and reproduction.

Regulating Gene Editing in Reproduction

To address these ethical concerns, it is necessary to establish comprehensive regulations on gene editing in reproduction. These regulations should strike a balance between allowing for advancements in science and technology while also ensuring the protection of individual and societal interests.

One approach is to prohibit the use of gene editing for non-medical purposes that could lead to the creation of "designer babies." However, this raises questions about defining what constitutes a medical necessity and the potential for subjective interpretations.

Another approach is to allow for gene editing only in cases where it can prevent the transmission of severe genetic diseases. This would require careful consideration of which conditions should be included and the potential implications for individuals carrying these conditions.

Transparency and public engagement are also crucial aspects of regulating gene editing in reproduction. Given the potential for wide-ranging societal impacts, including diverse perspectives in decision-making processes can help ensure more comprehensive and ethically sound policies.

Case Study: CRISPR Babies

The birth of the first genetically edited babies in China in 2018, commonly known as the "CRISPR babies" case, highlighted the urgency of addressing ethical concerns in gene editing. The scientist responsible for the experiment claimed to have used CRISPR-Cas9 to modify the genes of embryos to make them resistant to HIV infection. This controversial experiment sparked international outcry and led to calls for stricter regulations on gene editing research.

The case of the CRISPR babies illustrates the need for clear guidelines and oversight to prevent unethical practices and ensure responsible research conduct. It also underscores the importance of open dialogue and international cooperation to establish global standards for the ethical use of gene editing in reproduction.

Conclusion

Gene editing in reproduction has the potential to revolutionize healthcare and shape the future of humanity. However, it also raises profound ethical questions that must be carefully considered and addressed before widespread implementation.

Regulating gene editing in reproduction requires striking a delicate balance between scientific advancement, individual autonomy, and societal interests. It is essential to establish comprehensive regulations that safeguard against unethical practices while promoting equal access and responsible research conduct.

The future of reproduction lies in the ethical and responsible use of gene editing technologies, guided by principles of justice, beneficence, and respect for human dignity. As we navigate this uncharted territory, it is crucial to engage in ongoing discussions, considering diverse perspectives to shape a more equitable and ethically just future for reproductive health.

Ethical Considerations in Mental Health and Wellness

Understanding Mental Health Stigma

Historical Context of Mental Health Stigma

Understanding the historical context of mental health stigma is essential in comprehending the current state of mental health and the barriers to seeking help and support. Throughout history, individuals with mental health issues have been subjected to discrimination, misunderstanding, and social exclusion. This section will delve into the historical factors that have contributed to the development and perpetuation of mental health stigma.

Early Beliefs and Supernatural Explanations

Throughout ancient civilizations, mental health conditions were often attributed to supernatural forces or divine punishment. People with mental health issues were thought to be possessed by evil spirits, cursed, or even considered witches. These beliefs perpetuated fear and misunderstanding, resulting in the isolation and mistreatment of individuals with mental illnesses.

Religious and Philosophical Influences

Religious and philosophical beliefs have played a significant role in shaping attitudes towards mental health. In medieval Europe, mental illnesses were associated with moral failings and sin. The mentally ill were believed to be morally corrupt or possessed by demons, leading to harsh treatment, including exorcisms or confinement in asylums.

During the Enlightenment period, the concept of rationality gained prominence, and mental illness was often viewed as a deviation from reason. This view further marginalized those with mental health conditions, as they were seen as incapable of contributing to society.

The Rise of Asylums and Institutionalization

In the 18th and 19th centuries, the establishment of asylums marked a significant shift in the treatment of individuals with mental health conditions. Initially, asylums were intended to provide compassionate care and treatment. However, overcrowding, poor living conditions, and a lack of understanding about mental health led to deplorable situations. Patients were often subjected to inhumane treatments, including restraints and physical abuse.

The widespread use of institutionalization and the dehumanizing conditions in asylums contributed to the perpetuation of stigma. Society viewed mental illness as a shameful and incurable condition, leading to the marginalization and isolation of those suffering from mental health issues.

Emergence of Psychoanalysis and Medicalization of Mental Illness

The late 19th and early 20th centuries witnessed the emergence of psychoanalysis, pioneered by Sigmund Freud. Freud's theories provided a new framework for understanding mental health and emphasized the role of the unconscious mind. However, despite the advancements in understanding mental health, attitudes towards mental illness remained stigmatized.

The medicalization of mental illness reinforced the notion that mental health conditions were solely biological in nature. This perspective further segregated individuals with mental health issues, as they were perceived as flawed and different from the general population.

Media Influence and Stereotypes

The media, including films, literature, and news outlets, have played a significant role in perpetuating stereotypes and stigma surrounding mental health. Portrayals of individuals with mental health conditions as violent, unpredictable, or simply "crazy" have contributed to public fear and misunderstanding. Such representations reinforce societal prejudices and discourage individuals from seeking help due to fear of judgment and discrimination.

Advancements in Psychiatry and Deinstitutionalization

The latter half of the 20th century witnessed advancements in psychiatric treatments and the movement towards deinstitutionalization. The introduction of psychotropic medications and the development of psychotherapeutic approaches improved the ability to manage and treat mental health conditions.

However, the closure of many psychiatric hospitals without adequate community-based care led to significant challenges. Homelessness, incarceration, and inadequate support systems for individuals with mental illness became prevalent issues. These challenges further deepened the stigma and perpetuated misconceptions about mental health.

Current Efforts to Reduce Stigma

In recent years, there has been a growing recognition of the need to address mental health stigma and promote greater understanding and acceptance. Efforts from mental health organizations, advocacy groups, and public figures have aimed to challenge stereotypes, increase education, and promote empathy towards individuals with mental health conditions.

Anti-stigma campaigns, public awareness initiatives, and the sharing of personal stories have been instrumental in reducing mental health stigma. The focus has shifted towards viewing mental health as a part of overall well-being and promoting a more inclusive society that supports individuals with mental health conditions.

Conclusion

The historical context of mental health stigma reveals the deep-rooted biases and misunderstandings that have plagued society for centuries. Understanding this history is crucial in moving towards a more compassionate and inclusive future. Efforts to reduce mental health stigma should involve education, empathy, and a commitment to challenging societal norms that perpetuate discrimination. By addressing mental health stigma, we can create a more supportive environment and empower individuals to seek the help they need for their mental well-being.

Impact of Stigma on Mental Health Treatment

Stigma surrounding mental health is a significant barrier to accessing and receiving adequate treatment. It often leads to discrimination, lack of understanding, and negative attitudes toward individuals with mental health conditions. This stigma can have a profound impact on mental health treatment at both individual and societal levels. In this section, we will explore the various ways stigma affects mental health treatment and discuss strategies to address this issue.

Understanding Stigma

Stigma refers to the social disapproval and discrimination experienced by individuals who are perceived to deviate from societal norms. In the context of mental health, stigma can manifest as prejudice, stereotypes, and exclusion. This can lead to individuals being labeled as "crazy," "weak," or "dangerous," further reinforcing negative perceptions.

Stigma often arises due to a lack of awareness and education about mental health conditions. Many people have misconceptions and myths surrounding mental illnesses, which perpetuate fear and discrimination. Additionally, societal attitudes influenced by cultural, religious, and historical factors contribute to the stigmatization of mental health issues.

Impact on Help-Seeking Behavior

One of the major consequences of mental health stigma is that it discourages individuals from seeking help. People may fear judgment or rejection from their family, friends, or colleagues if they disclose their mental health struggles. The fear of being labeled as "crazy" or "unstable" often prevents individuals from speaking openly about their experiences and seeking appropriate treatment.

This reluctance to seek help can have severe consequences on individuals' well-being. Mental health conditions, when left untreated, can worsen and have long-term effects on an individual's overall functioning and quality of life. Stigma creates a barrier to timely intervention, resulting in delayed treatment and a higher risk of developing more serious mental health issues.

Quality and Effectiveness of Treatment

Stigma can also impact the quality and effectiveness of mental health treatment. Mental health professionals may hold stigmatizing beliefs or biases, which can affect their interactions with patients. These biases can result in inadequate care, misdiagnosis, and inappropriate treatment recommendations.

Furthermore, stigma can undermine the therapeutic relationship between the individual and the mental health professional. When individuals anticipate judgment or discrimination, they may be less likely to disclose sensitive information or fully engage in the treatment process. This can significantly hinder the therapeutic progress and outcome.

Societal Challenges

Stigma surrounding mental health extends beyond individual perceptions and attitudes. It can also impact public policies, funding allocations, and resource availability for mental health services. The negative stereotypes associated with mental illnesses often result in reduced support for mental health programs and a lack of investment in mental health infrastructure.

Moreover, stigma contributes to social exclusion and discrimination, limiting opportunities for individuals with mental health conditions to participate fully in society. This can lead to social isolation, unemployment, and compromised overall well-being.

Addressing Stigma in Mental Health Treatment

Reducing mental health stigma is crucial to ensure equitable access to quality mental health treatment. Several strategies can be employed to address this issue:

1. **Education and Awareness:** Promote education and awareness campaigns to debunk myths, provide accurate information about mental health, and promote empathy and understanding.

2. **Language and Media:** Encourage the use of respectful and non-stigmatizing language in media portrayal of mental health issues. Responsible reporting and challenging negative stereotypes can have a significant impact on public perception.

3. **Advocacy and Support:** Advocate for policies and legislation that protect the rights of individuals with mental health conditions, promote equal opportunities, and prevent discrimination.

4. **Shared Stories and Personal Narratives:** Encourage individuals with lived experiences of mental health conditions to share their stories to increase empathy and reduce stigma.

5. **Mental Health Promotion:** Invest in mental health promotion programs that foster resilience, well-being, and promote early intervention.

6. Collaboration and Partnerships: Foster partnerships between mental health organizations, healthcare professionals, community leaders, and policymakers to address stigma holistically.

Case Study: Let's Talk Campaign - India

One example of an initiative aimed at reducing stigma surrounding mental health is the "Let's Talk" campaign in India. This campaign, launched by the Indian government, seeks to raise awareness about mental health, reduce stigma, and encourage help-seeking behavior.

The Let's Talk campaign utilizes various strategies, including public service announcements, community events, and celebrity endorsements to reach a wide audience. It emphasizes the importance of open conversations about mental health and provides information on available support services.

The campaign has had a significant impact on public perceptions of mental health in India. It has led to increased awareness, decreased stigma, and a rise in the number of individuals seeking mental health services. The Let's Talk campaign demonstrates the power of education and awareness in challenging stigma and promoting mental health.

Conclusion

Stigma surrounding mental health has far-reaching effects on mental health treatment at both individual and societal levels. It discourages help-seeking behavior, compromises the quality of treatment, and perpetuates social exclusion. However, through education, awareness, advocacy, and collaboration, we can work towards reducing stigma and creating a more inclusive and supportive environment for individuals with mental health conditions. By addressing stigma, we pave the way for equitable access to quality mental health treatment and support the overall well-being of individuals and communities.

Note: The Let's Talk campaign is a real initiative in India, but the details provided in the case study are fictional for illustrative purposes.

Strategies for Reducing Mental Health Stigma

Mental health stigma refers to the negative attitudes, beliefs, and stereotypes that surround mental health conditions. It is a significant barrier to seeking help and support, as it contributes to feelings of shame, fear, and isolation among individuals experiencing mental health issues. Addressing mental health stigma is crucial for promoting the well-being of individuals and creating a more inclusive and supportive society. In this section, we will explore various strategies for reducing mental health stigma and creating a culture of acceptance and understanding.

Educating the Public

One of the most effective ways to combat mental health stigma is through education. It is important to provide accurate information about mental health conditions to dispel misconceptions and myths. Educational campaigns can use various channels such as schools, workplaces, community centers, and media platforms to increase awareness and understanding.

Educational initiatives should focus on:

+ Providing information about different mental health conditions, their causes, symptoms, and treatment options.

+ Highlighting the prevalence of mental health issues to demonstrate that they are common and can affect anyone.

+ Promoting the scientific basis of mental health conditions to challenge the notion that they are a result of personal weakness or character flaws.

+ Encouraging open discussions about mental health to create a safe and supportive environment where individuals feel comfortable sharing their experiences.

By providing accurate information and fostering discussions, education can help debunk myths, correct misconceptions, and reduce the fear and stigma associated with mental health.

Challenging Stereotypes

Stereotypes and stigmatizing language surrounding mental health contribute to the marginalization of individuals with mental health conditions. Challenging these stereotypes is an essential step in reducing mental health stigma. Here are some strategies to challenge stereotypes:

+ Promote Positive Portrayals: Media and entertainment play a powerful role in shaping public attitudes. Encouraging the portrayal of mental health in a positive and accurate light can help challenge stereotypes. This can be achieved by supporting movies, TV shows, and documentaries that depict mental health issues sensitively and realistically.

+ Language Matters: The use of stigmatizing language perpetuates negative stereotypes. Using person-first language (e.g., "person with schizophrenia" instead of "schizophrenic") emphasizes the individual's humanity rather than defining them by their condition. It is important to promote the use of inclusive language in all settings, including healthcare, media, and public discourse.

+ Sharing Personal Stories: Personal narratives have the power to humanize the experiences of individuals living with mental health conditions. Encouraging individuals to share their stories can help challenge stereotypes and foster empathy and understanding.

By challenging stereotypes and promoting positive portrayals, we can create a more compassionate and accepting society for individuals with mental health conditions.

Supporting Mental Health Services and Access

Access to affordable and quality mental health services is crucial for individuals experiencing mental health issues. Lack of access can further perpetuate stigma by sending the message that mental health is not a priority. To reduce mental health stigma, it is essential to support mental health services and ensure equitable access for all.

Here are some strategies to support mental health services and access:

+ Increase Funding: Adequate funding is critical to enhance the availability and quality of mental health services. Governments, policymakers, and organizations should prioritize mental health funding to expand resources, improve infrastructure, and train mental health professionals.

+ Integration of Services: Integrating mental health services into primary care settings, schools, workplaces, and community centers can contribute to reducing stigma. This approach normalizes mental health support and emphasizes the fact that mental health is an integral part of overall well-being.

+ Improve Affordability: High costs associated with mental health services can create significant barriers to access. Implementing policies to make mental health services affordable, such as insurance coverage for mental health treatments and subsidies for low-income individuals, can help ensure equitable access for all.

+ Enhance Mental Health Literacy: Promoting mental health literacy among individuals can facilitate early intervention and help-seeking behaviors. Providing education and resources on recognizing the signs of mental health issues and accessing appropriate support can encourage individuals to seek help without fear of judgment or stigma.

By supporting mental health services and access, we can create a society that values mental well-being and reduces the stigma associated with seeking help.

Promoting Peer Support and Community Engagement

Peer support and community engagement play a crucial role in reducing mental health stigma. Connecting individuals with shared experiences can foster a sense of belonging and reduce the isolation often associated with mental health conditions. Here are some strategies to promote peer support and community engagement:

+ Support Peer-Led Initiatives: Peer-led support groups, helplines, and online communities provide opportunities for individuals to connect with others who have similar experiences. These initiatives create safe spaces for sharing, support, and validation, reducing the stigma of mental health conditions.

+ Engage Community Leaders: Collaboration with community leaders, including religious leaders, teachers, and local influencers, can help facilitate conversations around mental health in various community settings. By engaging opinion leaders, we can promote acceptance and understanding within different cultural contexts.

+ Involve Schools and Colleges: Educational institutions play a crucial role in shaping attitudes and behaviors. By integrating mental health education into school curricula and providing support services, we can promote a culture of empathy and reduce stigma from a young age.

+ Workplace Mental Health Programs: Employers have a responsibility to create supportive and inclusive work environments. Implementing mental health programs, providing employee assistance programs, and training managers in recognizing and addressing mental health issues can reduce stigma in the workplace.

By promoting peer support, community engagement, and workplace initiatives, we can create communities that are empathetic, supportive, and free from mental health stigma.

Conclusion

Reducing mental health stigma is an ongoing process that requires collective effort from individuals, communities, organizations, and policymakers. By implementing strategies like education, challenging stereotypes, supporting mental health services, and promoting peer support, we can create a society that values mental health and well-being.

Remember, reducing mental health stigma is not just about improving the lives of individuals with mental health conditions but also about building a more compassionate, inclusive, and equitable society for everyone. Let us work together towards a future where mental health is fully understood, accepted, and supported.

Ethical Issues in Psychiatric Diagnosis and Treatment

Controversies in Psychiatric Diagnosis

Psychiatric diagnosis is a fundamental aspect of mental health care, providing clinicians with a tool for understanding and categorizing a patient's symptoms. However, the process of psychiatric diagnosis is not without its controversies. In this section, we will explore some of the key debates and challenges surrounding psychiatric diagnosis.

The Validity and Reliability of Diagnostic Categories

One of the main controversies in psychiatric diagnosis relates to the validity and reliability of diagnostic categories. Critics argue that the current diagnostic framework, as outlined in diagnostic manuals like the DSM-5 (Diagnostic and Statistical Manual of Mental Disorders), lacks scientific rigor and may not accurately capture the complexity of mental health conditions. They argue that diagnostic categories are often based on subjective clinical observations and lack clear diagnostic boundaries.

Moreover, the reliability of psychiatric diagnosis has also been questioned. Studies have shown variability in diagnostic decisions, indicating that different clinicians may diagnose the same patient with different disorders. This lack of consistency raises concerns about the reliability and consistency of psychiatric diagnoses.

The Medicalization of Normal Human Experiences

Another controversy in psychiatric diagnosis is the medicalization of normal human experiences. Critics argue that psychiatric diagnoses pathologize certain behaviors and emotions that may be within the normal range. For instance, the inclusion of conditions like "oppositional defiant disorder" and "disruptive mood dysregulation disorder" has been criticized for diagnosing children and adolescents with mental disorders for age-appropriate behaviors.

This controversy raises questions about the boundaries between "normal" and "abnormal" behavior and whether certain human experiences are being medicalized for the benefit of the pharmaceutical industry or other external influences.

Cultural and Contextual Considerations

Psychiatric diagnosis is often influenced by cultural and contextual factors, which can lead to controversies and challenges. For instance, certain symptoms and disorders may be more prevalent in specific cultural groups, leading to overdiagnosis or underdiagnosis based on cultural differences in symptom presentation.

Additionally, language barriers, cultural norms, and stigma surrounding mental health may affect the accuracy of psychiatric diagnosis in some populations. It is crucial for clinicians to be aware of these cultural and contextual considerations to ensure accurate and culturally sensitive diagnosis.

Comorbidity and Overlapping Symptomatology

Comorbidity refers to the presence of two or more mental health disorders in an individual. This raises challenges in psychiatric diagnosis, as many disorders share overlapping symptoms. For example, symptoms of anxiety and depression can coexist, making it difficult to distinguish between the two disorders.

The high rates of comorbidity in psychiatric disorders can complicate diagnosis and treatment planning. It requires clinicians to consider the underlying causes of the symptoms and exercise caution in labeling individuals with multiple diagnoses.

Alternative Paradigms and Frameworks

As the controversies surrounding psychiatric diagnosis continue, alternative paradigms and frameworks have emerged. One such approach is the dimensional model, which suggests that mental health conditions should be understood on a spectrum rather than discrete categories. This model emphasizes a person-centered approach, focusing on individual experiences rather than rigid diagnostic criteria.

Another alternative is the Recovery-Oriented Model, which emphasizes the personal and social aspects of recovery from mental health conditions. This model aims to empower individuals to take control of their lives and move beyond a diagnostic label.

Real-World Example: Attention-Deficit/Hyperactivity Disorder (ADHD)

ADHD is a commonly diagnosed neurodevelopmental disorder that has generated significant controversy. Critics argue that the diagnostic criteria for ADHD are vague and open to interpretation, leading to overdiagnosis, especially in children. The use of stimulant medications as a first-line treatment for ADHD has also raised concerns about overprescribing and potential long-term effects.

However, proponents of the diagnosis argue that ADHD is a valid and reliable condition, supported by extensive research. They highlight the benefits of early intervention and appropriate treatment in improving individuals' quality of life.

Conclusion

The controversies surrounding psychiatric diagnosis reflect the ongoing debates within the field of mental health. As our understanding of mental health continues to evolve, it is essential to critically examine the validity and reliability of diagnostic categories. By considering cultural and contextual factors, adopting alternative paradigms, and engaging in ongoing dialogue, we can strive for more nuanced and patient-centered approaches to psychiatric diagnosis.

Involuntary Commitment and Patient Rights

Involuntary commitment refers to the legal process by which individuals with mental health conditions can be admitted to a psychiatric hospital or facility against their will. This process often raises significant ethical concerns, as it involves restricting an individual's freedom and autonomy in order to protect their own safety or the safety of others. In this section, we will explore the ethical issues surrounding involuntary commitment and discuss the importance of protecting patient rights in psychiatric care.

Background and Legal Framework

Involuntary commitment laws vary across different jurisdictions, but they generally outline the circumstances under which an individual can be involuntarily admitted to a psychiatric facility. These laws typically require that the person poses a danger to themselves or others, is unable to care for themselves, or is unable to consent to treatment due to their mental health condition.

The goal of involuntary commitment is to provide immediate care and treatment for individuals who may be experiencing a mental health crisis. However, it is important to ensure that the process is conducted in a fair and ethical manner, with proper consideration for the individual's rights, autonomy, and dignity.

Ethical Considerations

Involuntary commitment raises several ethical concerns that need to be carefully considered. These include:

1. **Balancing autonomy and beneficence:** Involuntary commitment involves a tension between respecting an individual's autonomy and promoting their overall well-being. On one hand, individuals have the right to make decisions about their own healthcare and treatment. On the other hand, there may be situations where involuntary commitment is necessary to prevent harm to themselves or others. Striking the right balance between these competing principles is crucial.

2. **Avoiding stigmatization and discrimination:** Involuntary commitment can contribute to the social stigma surrounding mental illness. It is essential to ensure that individuals who are involuntarily committed are treated with dignity and respect, and that their rights are protected. Mental health conditions should not be a basis for prejudice or discrimination.

3. **Ensuring due process:** Involuntary commitment should be subject to appropriate legal safeguards and due process. This includes the right to legal representation, the right to challenge the commitment decision, and regular judicial review of the continued need for involuntary treatment. These safeguards help protect individuals from potential abuse or unjustified restrictions of their liberty.

4. **Promoting alternatives to involuntary commitment:** Whenever possible, efforts should be made to provide less restrictive alternatives to involuntary commitment. This may include crisis intervention services, community-based mental health treatment, or outpatient programs. Involuntary commitment should be seen as a last resort when all other options have been exhausted.

Case Study: Involuntary Commitment and Patient Rights

Let's consider the case of Sarah, a 32-year-old woman with a history of severe depression and suicidal ideation. Sarah has been voluntarily seeking treatment for her depression, but her condition has worsened recently, and she has expressed suicidal thoughts to her therapist.

Her therapist is concerned about Sarah's safety and believes that an involuntary commitment is necessary to ensure she receives immediate care and protection. However, Sarah is adamant that she should be allowed to make her own decisions regarding her treatment, and she refuses any form of hospitalization.

In this case, several ethical considerations come into play. Sarah's autonomy and right to make decisions about her treatment should be respected. However, her therapist also has a duty to prevent harm and ensure her safety. It may be necessary for the therapist to involve other healthcare professionals, such as a psychiatrist, to conduct a thorough assessment of Sarah's condition and explore alternative treatment options.

If it is determined that involuntary commitment is indeed necessary, the therapist should ensure that due process is followed, including involving legal professionals and providing Sarah with access to legal representation. Regular reviews of Sarah's condition and treatment should be conducted to reassess the need for continued involuntary commitment.

Ultimately, the key ethical principle guiding this situation should be the promotion of Sarah's well-being while balancing her autonomy and right to make decisions about her treatment. Collaboration, compassion, and respect for patient rights should guide the decision-making process in cases of involuntary commitment.

Resources and Further Reading

1. Appelbaum, P. S. (2010). Assessment of patient competence to consent to treatment. The New England Journal of Medicine, 357(18), 1834-1840.

2. Beauchamp, T. L., & Childress, J. F. (2019). Principles of biomedical ethics. Oxford University Press.

3. Gostin, L. O. (2020). Public health law: Power, duty, restraint. University of California Press.

4. Health and Human Rights Info. Involuntary Commitment. Retrieved from https://www.hhri.org/issues/mental-health/involuntary-commitment/

5. Radden, J. (2020). What is mental illness? Harvard University Press.

Exercise: Think about a hypothetical scenario where you are a mental health professional faced with the decision of whether to involuntarily commit a patient. Consider the ethical implications and discuss the different factors you would consider before making a decision. How would you approach balancing autonomy and the duty to prevent harm?

Psychopharmacology and Ethical Considerations

In the field of mental health, psychopharmacology plays a crucial role in the treatment of psychiatric disorders. Psychopharmacology is the study of how drugs affect the brain and behavior, and it encompasses the use of medications to manage symptoms and improve overall well-being. While psychopharmacology has revolutionized the treatment of mental illness, it also raises important ethical considerations that need to be carefully addressed.

The Role of Psychopharmacology in Mental Health Treatment

Psychopharmacology involves the use of medications to alter brain chemistry and improve symptoms related to psychiatric disorders. It aims to target specific neurotransmitters or receptor systems in the brain to restore the balance of chemicals that are associated with mental health conditions. Medications used in psychopharmacology include antidepressants, antipsychotics, mood stabilizers, anxiolytics, and stimulants.

One of the primary ethical considerations in psychopharmacology is the balance between the benefits and risks of medication. While medications can be highly effective in managing symptoms, they may also have adverse effects and potential risks. Medical professionals must carefully consider the potential benefits and risks of medication for each individual patient, taking into account their unique circumstances, medical history, and preferences.

Informed Consent and the Ethical Responsibility of the Prescriber

Informed consent is a fundamental ethical principle in healthcare that ensures patients have all the necessary information to make decisions about their treatment. When prescribing psychotropic medications, healthcare providers have an ethical responsibility to ensure that patients fully understand the potential benefits, risks, and alternatives.

In the context of psychopharmacology, informed consent should include a comprehensive discussion of the medication's intended effects, potential side effects, and possible drug interactions. The prescriber should also explain any required monitoring, dosage instructions, and the expected duration of treatment. In complex cases, it may be necessary to involve the patient's family or support network to ensure that decisions are made in the patient's best interest.

Ethical considerations also extend to the off-label use of medications. Off-label prescribing refers to the use of a medication for a purpose not approved by regulatory authorities. While off-label use can be appropriate in certain circumstances, prescribers must ensure that patients are fully informed about the potential risks and benefits, and that there is sufficient evidence to support the off-label use.

Balancing Personal Autonomy and Coercion

Another ethical challenge in psychopharmacology arises in situations where patients may lack the capacity to make informed decisions about their treatment. This can include individuals with severe mental illness, cognitive impairments, or those who are subjected to involuntary treatment. In such cases, balancing personal autonomy and the duty to provide care can be quite challenging.

Involuntary treatment, which involves administering medication without the patient's consent, raises ethical concerns regarding the potential infringement of individual rights. Healthcare professionals must navigate the delicate balance of providing necessary treatment while respecting an individual's autonomy. Legal frameworks and guidelines exist to regulate the circumstances under which involuntary treatment is permissible, with the aim of protecting the rights of patients.

Ethical Implications of Psychotropic Medications in Vulnerable Populations

Special attention must be given to the ethical implications of psychotropic medication use in vulnerable populations, such as children, pregnant women, and the elderly. These populations may be at higher risk of adverse effects or may have unique needs that require careful consideration.

When prescribing psychotropic medications to children, ethical concerns arise due to the long-term effects of medication on developing brains and bodies. The use of psychotropic medications in children should be based on thorough evaluation, consideration of non-pharmacological interventions, and a careful assessment of the potential risks and benefits.

Similarly, the use of psychotropic medications during pregnancy raises ethical considerations due to the potential effects on the developing fetus. Healthcare providers need to carefully balance the mental health needs of pregnant individuals with the potential risks to both the mother and the unborn child.

Elderly individuals may also be more susceptible to adverse effects from psychotropic medications due to age-related changes in metabolism and increased vulnerability to drug interactions. Prescribers must consider the unique needs and risks associated with medication use in the elderly population, including the potential for cognitive impairment and increased fall risks.

Promoting Ethical Practices in Psychopharmacology

To ensure ethical practices in psychopharmacology, it is essential to promote ongoing education and professional development for healthcare providers. This includes staying updated on the latest research findings, guidelines, and ethical principles in the field.

Open communication and collaboration among healthcare professionals, patients, and their support networks are vital in addressing ethical considerations. Shared decision-making processes that involve patients in treatment decisions, provide them with relevant information, and respect their autonomy can lead to more ethical psychopharmacological practices.

Additionally, research in psychopharmacology should prioritize ethical considerations, such as the inclusion of diverse populations, informed consent, privacy protection, and the appropriate use of placebos or control groups in clinical trials.

Conclusion

Psychopharmacology has revolutionized the field of mental health treatment, offering hope and relief to millions of individuals worldwide. However, it also brings forth various ethical considerations that need careful attention. By prioritizing informed consent, respecting personal autonomy, considering

vulnerabilities in certain populations, and promoting ethical practices, healthcare professionals can ensure the responsible and ethical use of psychotropic medications. Ultimately, an ethical approach to psychopharmacology contributes to the well-being and dignity of individuals seeking mental health treatment.

Mental Health and Wellness in Vulnerable Populations

Mental Health in Children and Adolescents

Mental health issues among children and adolescents have become a growing concern in recent years. The importance of addressing these issues cannot be understated, as they can have significant short-term and long-term effects on a child's well-being and development. In this section, we will explore the unique ethical considerations surrounding mental health in this population, as well as strategies for prevention, early intervention, and support.

Understanding Mental Health in Children and Adolescents

Mental health is a state of well-being in which an individual realizes their own potential, can cope with the normal stresses of life, and can work productively and fruitfully. In children and adolescents, mental health encompasses emotional, psychological, and social well-being, and is critical for their healthy development.

However, children and adolescents may experience a range of mental health issues, including anxiety disorders, mood disorders (such as depression), attention-deficit/hyperactivity disorder (ADHD), and eating disorders, among others. These issues can significantly impact their daily functioning, school performance, and overall quality of life.

It is essential to recognize that mental health issues can arise at any age, and early detection and intervention are crucial. Children and adolescents often face unique challenges and stressors, such as academic pressure, social expectations, peer pressure, and hormonal changes. It is therefore crucial to create a supportive environment that promotes mental well-being and provides appropriate interventions when necessary.

Ethical Considerations in Mental Health Treatment

When addressing mental health in children and adolescents, ethical considerations play a vital role in ensuring that their rights and well-being are protected. Here are some key ethical considerations to keep in mind:

1. **Informed Consent**   As children and adolescents may not have the capacity to make informed decisions about their mental health treatment, it is essential to involve their parents or legal guardians in the decision-making process. Informed consent from parents or guardians ensures that they are aware of the potential risks and benefits of the proposed treatment and have the opportunity to ask questions and seek alternative opinions. When appropriate, involving the child or adolescent in the decision-making process can also promote their autonomy and sense of agency.

2. **Confidentiality**   Confidentiality is a fundamental principle in mental health treatment. However, it is crucial to balance the need for confidentiality with the duty to protect children and adolescents from harm. Mental health professionals must navigate situations where the young person's well-being is at risk, such as when there are concerns of abuse, self-harm, or harm to others. In such cases, appropriate interventions, reporting, and collaboration with relevant authorities may be necessary to ensure the child's safety.

3. **Competency and Capacity Assessment**   Assessing the competency and capacity of children and adolescents is essential in determining their ability to make decisions about their mental health treatment. Mental health professionals should consider the child's developmental stage, cognitive abilities, and emotional maturity when assessing their competency. In cases where the child's capacity is in question, seeking additional input from parents, guardians, and other professionals can help inform the decision-making process.

4. **Cultural Competence**   Cultural competence is crucial in providing mental health support to children and adolescents from diverse backgrounds. Mental health professionals must be aware of and sensitive to cultural beliefs, practices, and values that may influence a child's understanding and acceptance of mental health treatment. Engaging with cultural communities and involving culturally competent professionals can enhance the effectiveness and relevance of mental health interventions.

Prevention, Early Intervention, and Support

Prevention and early intervention are key strategies in promoting mental health and addressing mental health issues in children and adolescents. Here are some approaches and resources that can support these efforts:

1. School-Based Programs   Schools play a vital role in promoting mental health among children and adolescents. Implementing school-based programs that focus on social-emotional learning, resilience-building, and mental health education can create a supportive environment for students. These programs can also include interventions such as counseling services, peer support groups, and mental health screenings to identify and address issues early on.

2. Parent and Caregiver Support   Providing education and support to parents and caregivers is crucial in promoting mental health and well-being in children and adolescents. Parenting programs that enhance parenting skills, communication, and coping mechanisms can contribute to a positive family environment. Additionally, providing access to resources, such as support groups and helplines, can help parents and caregivers navigate the challenges associated with their child's mental health.

3. Accessible and Affordable Mental Health Services   Ensuring access to high-quality, affordable mental health services is essential for children and adolescents. This includes adequate coverage and reimbursement for mental health services through insurance plans, reducing financial barriers to care. Community-based mental health services, including outpatient clinics, school-based health centers, and telehealth options, can improve access to care, particularly in underserved areas.

4. Peer Support and Mentoring   Engaging with peers who have had similar experiences can be a valuable source of support for children and adolescents. Peer support programs and mentoring initiatives can provide opportunities for young individuals to connect, share experiences, and learn coping strategies from their peers. These programs can be implemented within schools, community organizations, or online platforms.

5. Mental Health Education and Awareness Campaigns   Promoting mental health literacy among children, adolescents, and their communities is vital in reducing stigma and increasing help-seeking behavior. Educational campaigns can raise awareness about mental health issues, teach coping skills, and empower young individuals to seek support when needed. These campaigns can be organized through schools, community centers, and online platforms, using age-appropriate and culturally sensitive materials.

Case Study: The Impact of Bullying on Mental Health

Bullying is a prevalent issue in schools and can have severe detrimental effects on the mental health and well-being of children and adolescents. Let's consider a case study to understand the ethical considerations involved in addressing the mental health impact of bullying:

Emily, a 13-year-old girl, has been experiencing persistent bullying by her classmates at school. She is often subjected to verbal and physical abuse, which has led to feelings of worthlessness, withdrawal from social activities, and declining academic performance. Emily's parents are deeply concerned about her well-being and seek help from a mental health professional.

Ethical considerations in this case would include:

1. Informed consent: The mental health professional needs to obtain informed consent from Emily's parents or legal guardians regarding the proposed treatment plan and involvement of additional professionals if necessary.

2. Confidentiality: While maintaining confidentiality is important, the mental health professional may need to breach confidentiality if there is a risk of harm to Emily or if legal reporting obligations are involved.

3. Competency and capacity assessment: The mental health professional should assess Emily's ability to participate in her treatment decisions and consider her age, maturity level, and emotional capacity.

4. Cultural competence: Understanding Emily's cultural background and potential cultural factors related to bullying can help inform the treatment plan and ensure its effectiveness.

The mental health professional may adopt a multimodal approach to address Emily's mental health needs. This might involve individual therapy to address her emotional well-being, family therapy to help parents develop strategies to support her, and school-based interventions to address the bullying issue.

Additionally, raising awareness about bullying and its impact through school-wide campaigns, implementing anti-bullying policies, and providing training to teachers and students on recognizing and responding to bullying can contribute to the prevention and early intervention of such issues.

By addressing the mental health impact of bullying in an ethical and comprehensive manner, Emily's well-being can be restored, and her resilience and coping skills can be strengthened.

Summary

Mental health in children and adolescents is a multifaceted issue that requires careful consideration and ethical practice. By recognizing the unique challenges young individuals face and implementing preventive strategies, early interventions, and support systems, we can promote their mental well-being and overall quality of life. It is essential that mental health professionals, educators, parents, and communities work together to create a nurturing environment that fosters resilience and empowers young individuals to seek help when needed.

Mental Health in the Elderly

As individuals age, their mental health and well-being become increasingly important. The elderly population faces unique challenges that can significantly impact their mental health. In this section, we will explore the ethical considerations related to mental health in the elderly and the strategies to promote their well-being.

Understanding the Mental Health Challenges

The elderly population is at a higher risk of experiencing mental health disorders compared to younger age groups. Some common mental health challenges faced by the elderly include depression, anxiety, dementia, and social isolation.

Depression is one of the most prevalent mental health disorders in the elderly. It is often underdiagnosed and undertreated, leading to adverse consequences for the individual's overall well-being. Anxiety disorders, such as generalized anxiety disorder and phobias, can also significantly impact the daily lives of elderly individuals.

Dementia, including Alzheimer's disease, is a progressive neurodegenerative disorder characterized by cognitive decline and memory loss. It poses unique ethical challenges, particularly regarding decision-making capacity and the autonomy of the individual.

Social isolation and loneliness are also major concerns for the mental health of the elderly population. Lack of social support and limited social interactions can lead to feelings of isolation, depression, and anxiety.

Ethical Considerations

Ethical considerations play a crucial role in addressing mental health issues in the elderly population. The following ethical principles guide the provision of mental health care for the elderly:

Autonomy and Dignity: Respecting the autonomy and dignity of elderly individuals is paramount. They should have the right to make informed decisions regarding their mental health care and be treated with respect and empathy.

Beneficence and Non-Maleficence: Healthcare professionals should strive to promote the well-being of elderly individuals and prevent harm. This includes providing appropriate treatment and support to alleviate mental health symptoms and enhancing their quality of life.

Justice and Equity: Ensuring equitable access to mental health care is essential. Elderly individuals, regardless of their socioeconomic status or geographical location, should have access to appropriate mental health services.

Confidentiality: Respecting the confidentiality of elderly individuals' mental health information is crucial for building trust. Healthcare professionals must ensure that sensitive information is protected and shared only with the informed consent of the individual or as required by law.

Strategies for Promoting Mental Health in the Elderly

Promoting mental health in the elderly requires a comprehensive and multidimensional approach. The following strategies can be employed:

Early Detection and Screening: Regular mental health screenings can help identify early signs of mental health disorders in elderly individuals. Healthcare professionals should be trained to recognize and assess mental health symptoms in the elderly population.

Person-Centered Care: Taking a person-centered approach that recognizes the unique needs and preferences of each elderly individual is crucial. Mental health care should be tailored to their specific circumstances and incorporate their values and goals.

Integrated Care: Integrating mental health care into primary care settings can improve access and outcomes for the elderly population. Collaborative efforts between mental health professionals and primary care providers can enhance the detection and management of mental health conditions.

Social Support and Community Engagement: Encouraging social interactions and engagement in meaningful activities can combat social isolation and loneliness. Community-based programs, support groups, and volunteer opportunities can provide the necessary social support for the elderly.

Education and Training: Educating healthcare professionals, caregivers, and family members about mental health issues in the elderly is vital. Proper training can enhance awareness, reduce stigma, and improve the quality of care provided to elderly individuals.

Caregiver Support: Providing support and resources for caregivers of elderly individuals with mental health disorders is essential. Respite care, counseling, and education can help caregivers better manage the challenges associated with caring for the elderly with mental health conditions.

Case Study: Addressing Depression in Older Adults

Let's consider the case of Mr. Johnson, a 75-year-old man who has been exhibiting signs of depression. He has lost interest in activities he once enjoyed, experiences appetite changes, and has difficulty sleeping. Mr. Johnson lives alone and has limited social interactions.

To address Mr. Johnson's depression, a comprehensive approach is needed. His primary care physician conducts a mental health screening and diagnoses him with depression. The physician then collaborates with a mental health professional to develop a person-centered care plan.

The care plan includes a combination of psychotherapy and medication tailored to Mr. Johnson's needs. Additionally, the physician connects Mr. Johnson with local support groups for social engagement and arranges regular follow-up appointments to monitor his progress.

The physician also provides education and support to Mr. Johnson's family members, ensuring they understand the nature of depression and can provide appropriate support at home. The family is encouraged to participate in caregiver support programs to enhance their ability to care for Mr. Johnson's mental health.

By implementing this comprehensive approach, Mr. Johnson's depression symptoms improve gradually, and he experiences an enhanced sense of well-being and improved quality of life.

Conclusion

Addressing mental health challenges in the elderly requires a compassionate and ethical approach. By recognizing the unique mental health needs of the elderly population and employing person-centered care, integrated services, and community engagement, we can promote their mental well-being and ensure a dignified and fulfilling life in their later years.

Mental Health in Incarcerated Individuals

Mental health is a critical aspect of overall well-being that affects individuals from all walks of life, including those who are incarcerated. The prison system is known for its high prevalence of mental health issues, with studies suggesting that rates of mental disorders among incarcerated individuals are significantly higher compared to the general population. This section explores the ethical considerations surrounding mental health in incarcerated individuals, the challenges they face, and potential strategies for addressing these issues.

Understanding the Mental Health Needs of Incarcerated Individuals

The mental health needs of incarcerated individuals are complex and multifaceted. Many individuals enter the criminal justice system with pre-existing mental health conditions, such as depression, anxiety disorders, substance use disorders, or severe mental illnesses like schizophrenia or bipolar disorder. The stress and trauma of the prison environment can exacerbate these conditions and also contribute to the development of new mental health problems.

Incarcerated individuals often face a range of challenges that impact their mental well-being. These challenges may include exposure to violence, limited access to healthcare, overcrowded conditions, social isolation, and lack of continuity of care. Additionally, the stigma surrounding mental health within the prison system can deter individuals from seeking necessary treatment or support.

Ethical Considerations and Challenges

Addressing the mental health needs of incarcerated individuals raises a host of ethical considerations.

One such consideration is the principle of justice. It is crucial to ensure that incarcerated individuals have equitable access to mental health services that meet their unique needs. However, limited resources within the prison system can pose challenges in providing adequate mental health care to all individuals. Allocation of resources, prioritization of treatment, and ensuring fairness in access to care are ongoing ethical concerns.

Another consideration is the principle of autonomy. In order to promote the well-being of incarcerated individuals, it is essential to respect their autonomy and involve them in decision-making regarding their mental healthcare. However, the prison environment may restrict individuals' autonomy, making it necessary to navigate the ethical tension between ensuring their safety and respecting their autonomy.

The principle of beneficence also comes into play, as the aim should be to provide effective mental health interventions that promote the well-being of incarcerated individuals. This requires a commitment to evidence-based practices and ensuring that mental health professionals within the prison system receive proper training and support.

Addressing Mental Health in Incarcerated Individuals

To address the mental health needs of incarcerated individuals, a comprehensive and multidimensional approach is necessary. This approach should include:

1. **Screening and Assessment:** Implementing systematic screening and assessment procedures to identify mental health needs upon entry into the prison system. This helps in determining appropriate treatment plans and ensures that individuals receive timely and targeted interventions.

2. **Treatment and Support:** Providing evidence-based mental health treatment and support services within the prison system. This can include individual or group therapy, psychoeducation, psychotropic medication management, and addiction treatment programs. Collaborative efforts between mental health professionals, correctional staff, and community-based organizations are crucial for effective treatment.

3. **Continuity of Care:** Ensuring continuity of mental health care during transitions from the correctional facility to the community. This includes coordinating with community-based mental health providers to establish post-release treatment plans and facilitate access to resources upon release.

4. **Education and Training:** Providing education and training to correctional staff on recognizing and responding to mental health needs. This can help create a more supportive and understanding environment within the prison system.

5. **Peer Support Programs:** Implementing peer support programs where incarcerated individuals with lived experience of mental illness can provide support and guidance to their peers. Peer support has shown to be effective in reducing stigma, improving engagement in treatment, and promoting recovery.

6. **Reentry Support:** Offering comprehensive reentry support services that address housing, employment, and social support needs. These services can help reduce the risk of recidivism and support successful community integration.

Case Study: The Step Forward Program

The Step Forward Program is an example of an innovative approach to addressing mental health in incarcerated individuals. This program, implemented in a state correctional facility, focuses on providing evidence-based mental health treatment and support services. It emphasizes the importance of trauma-informed care, cognitive-behavioral therapy, and peer support.

The program incorporates both individual and group therapy sessions, focusing on building coping skills, improving emotional regulation, and addressing underlying trauma. Peer support specialists, who have personal experience with incarceration and mental illness, play a vital role in providing support and guidance to participants.

Evaluation of the Step Forward Program has shown promising results in reducing mental health symptoms, improving psychological well-being, and reducing disciplinary infractions within the correctional facility. The program has also demonstrated positive outcomes in terms of successful community reintegration upon release.

Conclusion

Addressing mental health in incarcerated individuals is a complex and ethically challenging endeavor. It requires a comprehensive approach that considers the unique needs of this population and navigates the ethical tensions surrounding justice, autonomy, and beneficence.

By implementing evidence-based practices, providing targeted interventions, and promoting continuity of care, it is possible to improve mental health outcomes for incarcerated individuals. This not only benefits the individuals themselves but also contributes to safer and more rehabilitative prison environments and better outcomes for communities as a whole.

However, addressing mental health in the prison system is not a standalone solution. Efforts must also be made to address broader issues such as systemic inequalities, social determinants of health, and the overreliance on incarceration as a response to societal problems. Only through a holistic, multifaceted approach can we strive towards a more just and equitable future for mental health in the context of incarceration.

Mental Health in the LGBTQ+ Community

Mental health is a crucial aspect of overall well-being, and it is influenced by several factors, including social, cultural, and individual experiences. In recent years, there has been growing recognition of the unique mental health challenges faced by individuals within the LGBTQ+ community. This subsection explores the specific issues affecting the mental health of LGBTQ+ individuals, the underlying causes, and potential strategies for support and intervention.

Understanding the LGBTQ+ Community

Before delving into the mental health challenges faced by the LGBTQ+ community, it is important to have a basic understanding of this diverse group. The LGBTQ+ acronym stands for lesbian, gay, bisexual, transgender, queer/questioning, and the '+' represents other sexual orientations and gender identities not explicitly mentioned. This inclusive terminology recognizes the broad spectrum of identities and experiences within this community.

Types of Mental Health Challenges

Members of the LGBTQ+ community often experience higher rates of mental health disorders compared to the general population. These challenges can arise due to a range of factors, including minority stress, discrimination, and internalized stigma. Some common mental health issues observed among LGBTQ+ individuals include:

- **Depression and anxiety:** LGBTQ+ individuals may experience chronic stress due to discrimination, rejection, or the fear of coming out. This can lead to higher rates of depression and anxiety disorders.

- **Suicidality:** LGBTQ+ individuals are at an increased risk of suicidal ideation and attempts. This vulnerability may stem from the lack of acceptance and social support, as well as the experience of discrimination and victimization.

+ **Substance abuse:** A higher prevalence of substance use disorders is observed among LGBTQ+ individuals, which can be linked to the use of substances as a coping mechanism for stress and discrimination.

+ **Eating disorders:** LGBTQ+ individuals may be more susceptible to developing eating disorders, such as anorexia nervosa or bulimia, as a result of body image concerns, internalized homophobia, or gender dysphoria.

Factors Contributing to Mental Health Challenges

Several factors contribute to the mental health challenges faced by LGBTQ+ individuals. These factors can interact and exacerbate each other, resulting in a greater impact on mental well-being. Here are some key contributing factors:

+ **Minority stress:** This refers to the chronic stress experienced by individuals from stigmatized groups due to their minority status. LGBTQ+ individuals face unique stressors, including coming out, family rejection, workplace discrimination, and societal prejudice. This constant exposure to stress can contribute to mental health difficulties.

+ **Lack of social support:** Support from family, friends, and communities plays a crucial role in mental health and well-being. LGBTQ+ individuals may face rejection from family members or feel isolated due to societal attitudes. The absence of supportive networks can increase the risk of mental health problems.

+ **Internalized stigma:** Internalized homophobia, biphobia, or transphobia refers to the self-negative attitudes and beliefs that individuals from the LGBTQ+ community may internalize due to societal biases. This self-stigmatization can lead to feelings of worthlessness, guilt, and shame, impacting mental health.

+ **Healthcare disparities:** LGBTQ+ individuals may encounter barriers to accessing appropriate healthcare, including mental health services. Discrimination and lack of provider knowledge around LGBTQ+ issues can result in inadequate care, delaying or impeding proper treatment.

+ **Intersectionality:** Intersectionality recognizes that individuals may experience overlapping forms of discrimination and disadvantage based on multiple aspects of their identity (e.g., race, ethnicity, gender, sexual orientation). LGBTQ+ individuals who belong to marginalized racial or ethnic groups may face compounded mental health challenges due to intersecting forms of oppression.

Strategies for Support and Intervention

To address the mental health needs of LGBTQ+ individuals, it is crucial to adopt a multidimensional and inclusive approach. Here are some strategies to consider:

- **Culturally competent care:** Mental health professionals should receive education and training to understand the unique challenges faced by LGBTQ+ individuals. This includes developing knowledge about sexual orientation, gender identity, and the impact of systemic discrimination on mental health.

- **Creating inclusive environments:** Educational institutions, workplaces, and healthcare settings should work towards fostering inclusive and affirming environments for LGBTQ+ individuals. This involves implementing policies that prohibit discrimination and promoting diversity, as well as providing resources and support services.

- **Social support networks:** Building strong social support networks is crucial for LGBTQ+ individuals' mental health. Creating safe spaces, support groups, and community organizations can offer a sense of belonging and reduce isolation.

- **Access to affirmative mental health services:** Mental health services should be tailored to meet the specific needs of LGBTQ+ individuals. Providing accessible, affordable, and culturally affirming therapy can help address mental health concerns effectively.

- **Advocacy and policy change:** Promoting social equality and fostering policies that protect the rights of LGBTQ+ individuals are essential for creating a more inclusive society. Advocacy efforts can help reduce discrimination, stigma, and improve access to mental healthcare.

Case Study: Mental Health Support for LGBTQ+ Youth

One real-world example of supporting mental health in the LGBTQ+ community is the Trevor Project, a leading organization providing crisis intervention and suicide prevention services to LGBTQ+ youth. With a 24/7 helpline, online chat, and text message support, they offer immediate assistance to individuals in crisis.

Additionally, they provide resources for mental health professionals and advocate for policies that protect LGBTQ+ youth.

Conclusion

The mental health challenges faced by LGBTQ+ individuals are multifaceted and influenced by social, cultural, and individual factors. Recognizing and addressing these challenges requires a comprehensive approach that includes education, advocacy, and the provision of accessible and affirming mental health services. By promoting acceptance, reducing stigma, and fostering inclusive environments, we can create a more supportive and equitable future for the mental health and wellness of the LGBTQ+ community.

Ethical Considerations in Psychotherapy and Counseling

Confidentiality and Informed Consent in Therapy

Confidentiality and informed consent are two essential ethical principles in the field of therapy. These principles guide the therapist's actions to ensure the privacy and autonomy of the client, while also promoting trust and a therapeutic relationship. In this section, we will explore the importance of confidentiality and informed consent in therapy, discuss the ethical considerations associated with these principles, and provide practical guidance for therapists.

Confidentiality in Therapy

Confidentiality refers to the obligation of therapists to keep the information shared by their clients confidential. The confidentiality of client information is crucial as it fosters a safe and non-judgmental space for clients to share their thoughts, feelings, and experiences.

As a therapist, it is essential to maintain strict confidentiality to promote trust and respect. Clients should feel confident that their personal information will not be disclosed without their explicit consent. Breaching confidentiality without a valid reason can harm the therapeutic relationship and deter individuals from seeking therapy.

However, there are certain situations when therapists may need to breach confidentiality. These situations include:

1. **Harm to self or others:** If the therapist believes that the client presents a serious risk of harm to themselves or others, they have a duty to notify the appropriate authorities. This duty of care takes precedence over maintaining confidentiality.

2. **Legal obligations:** Therapists may be legally required to disclose client information in certain circumstances, such as suspected child abuse or when subpoenaed by a court.

3. **Consultation and supervision:** Therapists often seek consultation and supervision to ensure the quality of their services. However, when discussing cases, therapists must protect their clients' identities and maintain confidentiality.

To ensure confidentiality, therapists should explain the limits of confidentiality in their initial sessions and obtain the client's written consent for treatment. Additionally, they should store client records securely and limit access to authorized personnel.

Informed Consent in Therapy

Informed consent is the process through which therapists obtain the client's voluntary agreement to receive treatment. It involves providing the client with comprehensive information about the nature of therapy, the therapist's qualifications, the proposed treatment plan, potential risks and benefits, and alternative options.

The informed consent process allows clients to make autonomous decisions about their mental health care. It ensures that clients have a clear understanding of what to expect from therapy and enables them to provide input into their treatment goals and methods. Informed consent also serves as a legal and ethical safeguard for both the client and the therapist.

When obtaining informed consent, therapists should:

+ Explain the purpose and goals of therapy in a language that the client can understand.

+ Discuss the therapist's qualifications, experience, and theoretical orientation.

+ Describe the proposed treatment plan, including the frequency and duration of sessions.

+ Outline the potential risks, benefits, and limitations of therapy.

+ Discuss any foreseeable conflicts of interest or dual relationships that may affect the therapeutic process.

+ Obtain the client's voluntary agreement to participate in therapy, ensuring they have the capacity to provide consent.

It is important to note that informed consent is an ongoing process throughout therapy. As therapy progresses, therapists should continuously keep clients informed about any changes in treatment, address any concerns or questions they may have, and obtain further consent for any adjustments to the treatment plan.

Ethical Considerations

Confidentiality and informed consent in therapy raise several ethical considerations that therapists must navigate. These considerations include balancing client autonomy with the duty to protect clients and others, addressing power imbalances in the therapeutic relationship, and ensuring cultural sensitivity and competence.

Therapists must respect the autonomy and privacy of their clients by upholding the principles of confidentiality and informed consent. However, they also have an ethical responsibility to safeguard clients and others from harm. This delicate balance requires therapists to carefully assess situations where breaching confidentiality may be necessary to protect the client or others from imminent danger.

Power imbalances inherent in the therapeutic relationship also need to be addressed. Clients may feel vulnerable and may not fully understand the implications of treatment or their rights. It is the therapist's responsibility to communicate clearly, check for understanding, and create a safe and empowering environment where clients feel comfortable asking questions and expressing their concerns.

Cultural sensitivity and competence are essential in confidentiality and informed consent in therapy. Therapists must recognize and respect cultural differences, actively engage in ongoing education and self-reflection, and adapt their practices to meet the unique needs of diverse clients. Informed consent should be obtained in a culturally appropriate manner, taking into consideration values, beliefs, and communication styles.

Case Study: Balancing Confidentiality and Safety

Consider the case of Sarah, a therapist working with a client named John who has been expressing thoughts of self-harm during their sessions. John explicitly asks Sarah not to share this information with anyone, emphasizing the importance of confidentiality to him.

As a therapist, Sarah faces a dilemma between honoring John's request for confidentiality and her duty to ensure his safety. In this situation, Sarah should prioritize the duty to protect John's life over maintaining strict confidentiality. Sarah should openly discuss her concerns with John, emphasizing the importance of his safety.

Sarah may explore alternative options with John, such as involving his support system or other mental health professionals, with his consent. However, if Sarah believes that John's risk of self-harm is immediate and serious, she should breach confidentiality and involve appropriate authorities to ensure his safety.

This case underscores the complexity of balancing confidentiality and safety. Therapists must carefully assess each situation, considering the client's autonomy and the potential harm that may arise from maintaining strict confidentiality.

Conclusion

Confidentiality and informed consent form the foundation of ethical practice in therapy. Therapists must uphold the privacy and autonomy of their clients while ensuring their safety and well-being. By navigating these principles ethically, therapists can foster a trusting therapeutic relationship and promote positive outcomes for their clients. It is essential for therapists to stay informed about the latest developments in ethical guidelines and seek consultation or supervision when facing ethical dilemmas.

Dual Relationships and Boundaries in Therapy

In the field of therapy, one of the most important ethical considerations is maintaining appropriate boundaries between therapists and their clients. This includes avoiding dual relationships, which occur when therapists have multiple roles or relationships with clients outside of the therapeutic setting. Dual relationships can potentially compromise the therapeutic alliance and the well-being of clients. In this section, we will explore the concept of dual relationships, the potential ethical issues they present, and strategies for managing and navigating these complex dynamics.

Understanding Dual Relationships

Dual relationships in therapy occur when therapists have simultaneous or consecutive roles with clients that extend beyond the therapeutic relationship. Examples of dual relationships include being a therapist and a close friend, a therapist and a family member, or a therapist and a business partner. These relationships can introduce conflicts of interest, power imbalances, and breaches of confidentiality.

The boundary between therapists and clients is crucial to maintain a safe and therapeutic environment. It allows clients to feel secure, trust their therapist, and share their deepest thoughts and emotions without fear of judgment or exploitation. When a dual relationship exists, the lines between the professional and personal become blurred, and the therapeutic relationship can be compromised.

Ethical Issues and Concerns

Dual relationships pose several ethical issues that therapists must be aware of and navigate carefully. Some of the key concerns include:

1. **Conflict of Interest:** When therapists have multiple roles with clients, conflicts of interest can arise. These conflicts may affect the therapist's ability to provide unbiased and objective treatment. For example, if a therapist is also a close friend of a client, their personal relationship may cloud their judgment or lead to biased decision-making in the therapeutic process.

2. **Power Imbalance:** Therapists hold a position of power and authority in the therapeutic relationship. Engaging in a dual relationship can exacerbate this power dynamic, potentially leading to exploitation or the misuse of this power. Clients may feel pressured to comply with the therapist's requests or disclose personal information outside the scope of therapy due to the power differential.

3. **Confidentiality and Privacy:** Dual relationships can also compromise the confidentiality and privacy of clients. If a therapist has multiple roles with a client, it becomes challenging to maintain appropriate boundaries and protect the client's personal information. Confidentiality breaches can occur if the therapist involuntarily discloses sensitive information in unrelated contexts.

4. **Impairment of Objectivity:** Dual relationships can impair a therapist's objectivity and professional judgment. It becomes difficult to separate

personal feelings or biases from the therapeutic process, potentially impacting treatment decisions and interventions. This impairment can compromise the quality and effectiveness of therapy.

Managing Dual Relationships

While complete avoidance of all dual relationships may not always be feasible or practical, therapists must carefully manage and navigate these complex dynamics to minimize ethical concerns. Here are some strategies for effectively handling dual relationships:

1. **Awareness and Self-Reflection:** Therapists must actively reflect on their own boundaries, values, and motivations when considering potential dual relationships. Self-awareness and ongoing self-reflection can help therapists identify potential conflicts of interest or power imbalances in their relationships with clients.

2. **Consultation and Supervision:** Seeking consultation and supervision from experienced professionals can provide valuable guidance and support when facing dilemmas related to dual relationships. Supervisors can help therapists gain different perspectives, identify potential ethical pitfalls, and explore alternative approaches to managing these relationships.

3. **Informed Consent:** When a situation arises where a dual relationship may be unavoidable, therapists should obtain informed consent from their clients. Informed consent involves discussing the potential risks and benefits of engaging in a dual relationship, ensuring that clients fully understand the implications and have the freedom to consent or decline without fear of repercussions.

4. **Establishing Clear Boundaries:** Therapists should establish and communicate clear boundaries with their clients from the outset of the therapeutic relationship. This includes discussing the limitations of the therapeutic relationship and outlining expectations regarding non-therapeutic interactions or relationships outside of therapy.

5. **Continued Monitoring and Evaluation:** Therapists should continuously monitor and evaluate the impact of any dual relationships on the therapeutic process. Regularly reassessing the boundaries and dynamics of the relationship helps ensure that the client's well-being remains the central focus and that any potential ethical concerns are promptly addressed.

Case Study: Managing a Dual Relationship

Consider the following case study to illustrate the complexities of managing a dual relationship:

Sarah is a therapist counseling a teenage client, Alex, who is struggling with anxiety and depression. Sarah is also a member of the same community organization as Alex's parents. The parents approach Sarah and request her assistance in planning a fundraising event for the organization. Sarah recognizes the potential dual relationship and the associated ethical concerns.

To effectively manage this situation, Sarah should:

1. Reflect on her motivations and potential conflicts of interest in engaging in the dual relationship.

2. Consult with her supervisor or a trusted colleague to gain an external perspective and explore alternative options.

3. Discuss the situation with Alex and their parents, ensuring informed consent and transparency about the potential impact on the therapeutic relationship.

4. Establish clear boundaries and expectations for maintaining client confidentiality and separating the dual relationship from therapy.

5. Continuously monitor the impact of the dual relationship on the therapeutic process and be prepared to make adjustments if any ethical concerns arise.

By following these steps, Sarah can navigate the dual relationship while prioritizing the well-being and best interests of her client, Alex.

Conclusion

Dual relationships pose significant ethical challenges in the field of therapy. Maintaining appropriate boundaries is crucial to ensuring the safety, trust, and well-being of clients. By understanding the potential ethical issues, therapists can navigate these complex dynamics with awareness, self-reflection, clear communication, and ongoing evaluation. Effective management of dual relationships strengthens the therapeutic alliance and upholds the ethical standards of the profession, thus promoting positive client outcomes and maintaining the integrity of the therapeutic process.

Cultural Competence in Therapy

Cultural competence is a crucial aspect of providing effective therapy to individuals from diverse backgrounds. It involves understanding and appreciating the influence of culture on a person's beliefs, values, behaviors, and mental health. In this section, we will explore the importance of cultural competence in therapy, discuss key concepts and principles, and provide strategies for developing cultural competence in therapeutic practices.

Understanding Cultural Competence

Cultural competence can be defined as the ability of healthcare professionals, including therapists, to effectively interact with individuals from diverse cultural backgrounds. It involves the integration of knowledge, awareness, and skills that enable therapists to provide culturally sensitive and responsive care.

In therapy, cultural competence acknowledges that individuals' experiences, values, and beliefs are shaped by their cultural contexts. It recognizes that culture influences how individuals perceive mental health, seek help, express their symptoms, and respond to treatment. By understanding and respecting these cultural factors, therapists can create a safe and supportive environment for their clients.

Key Principles of Cultural Competence

To develop cultural competence in therapy, therapists should adhere to several key principles:

1. **Self-awareness:** Therapists need to reflect on their own cultural biases, assumptions, and values. It's important to be aware of how their own cultural background may influence their perception of clients and their experiences.

2. **Knowledge of diverse cultures:** Therapists should actively seek knowledge about various cultures, including their beliefs, practices, and values. This knowledge helps to understand and appreciate the cultural context of their clients.

3. **Respect and empathy:** Therapists must approach each client with respect and empathy, valuing their unique experiences and perspectives. It is essential to create a non-judgmental and inclusive therapeutic space.

4. **Avoiding stereotypes and generalizations:** Therapists should be mindful of avoiding stereotypes and generalizations about clients based on their cultural backgrounds. Each individual is unique, and cultural factors should be considered in conjunction with other personal characteristics.

5. **Effective communication:** Therapists should develop effective communication skills that consider cultural nuances. This includes being aware of non-verbal cues, language barriers, and different communication styles.

6. **Collaboration and empowerment:** Therapists should prioritize collaboration and involve clients in the therapeutic process. By empowering clients to actively participate in treatment decisions, therapists can honor their autonomy and cultural values.

Strategies for Developing Cultural Competence

Developing cultural competence in therapy is an ongoing process that requires continuous learning and self-reflection. Therapists can adopt the following strategies to enhance their cultural competence:

1. **Continuing education and training:** Therapists should seek out educational opportunities to learn about diverse cultures, cultural competence, and cultural humility. This may include attending workshops, conferences, or cultural immersion experiences.

2. **Engaging with diverse communities:** Actively engaging with individuals from diverse cultures can foster understanding and appreciation of different perspectives. This can be done through community involvement, volunteering, or participating in cultural events.

3. **Supervision and consultation:** Seeking supervision or consultation from experienced therapists who have expertise in cultural competence can provide valuable guidance and feedback. Supervisors can help therapists navigate cultural challenges and expand their knowledge.

4. **Self-reflection and self-awareness:** Regularly reflecting on personal biases, assumptions, and reactions is essential for developing cultural competence. Therapists should actively engage in self-reflection and take steps to address any biases that may impact their therapeutic relationships.

5. **Culturally responsive assessment and treatment:** Therapists should adapt their assessment and treatment approaches to consider the cultural context of their clients. This may involve using culturally sensitive assessment tools, modifying treatment modalities, or integrating cultural practices in therapy.

6. **Seeking feedback from clients:** Engaging in open and honest communication with clients about their cultural needs and preferences can help therapists deliver more effective and culturally responsive care. Seeking feedback and adapting therapy accordingly demonstrates a commitment to cultural competence.

7. **Building diverse professional networks:** Collaborating with professionals from diverse cultural backgrounds can provide therapists with different perspectives and insights. Building a diverse network fosters cultural exchange and deepens understanding.

Case Example

To illustrate the importance of cultural competence in therapy, consider the following case:

Maria, a therapist, is working with a client, Ahmed, who comes from a Middle Eastern cultural background. Ahmed expresses concerns about seeking mental health treatment, as mental health is stigmatized in his community. Maria, recognizing the impact of culturally rooted stigma, creates a safe and non-judgmental environment for Ahmed to discuss his concerns. She takes the time to understand his cultural background, his values, and his reservations about seeking help. Maria collaborates with Ahmed to develop a treatment plan that respects his cultural values, integrates elements of his cultural background, and reduces the fear of being stigmatized.

This case demonstrates the importance of cultural competence in therapy. By recognizing and addressing Ahmed's cultural concerns, Maria can provide culturally sensitive care that meets Ahmed's unique needs.

Conclusion

Cultural competence is an essential aspect of therapy that promotes inclusivity, respect, and effectiveness. By understanding the influence of culture on individuals' mental health experiences, therapists can provide culturally sensitive and responsive care. Developing cultural competence requires ongoing self-reflection, knowledge acquisition, and the implementation of strategies that prioritize the diverse needs of clients. By embracing cultural competence, therapists can foster a therapeutic environment that supports individuals from all cultural backgrounds.

Ethical Issues in Online Counseling

Online counseling, also known as e-counseling or teletherapy, refers to the provision of mental health services through digital platforms such as video conferencing, text messaging, or email. This mode of counseling has gained popularity in recent years due to its convenience and accessibility. However, it presents unique ethical challenges that need to be carefully navigated to ensure the delivery of ethical, safe, and effective mental health care.

Confidentiality and Privacy Concerns

One of the primary ethical concerns in online counseling is maintaining confidentiality and privacy. In traditional face-to-face counseling, the therapist creates a private and secure space where clients can freely share their personal information without fear of it being disclosed to others. However, in the online setting, there is an increased risk of breaches in confidentiality and privacy due to potential technological vulnerabilities.

Therapists providing online counseling must ensure that the chosen communication platforms are secure and encrypted to protect client information. They should also inform clients about the limitations of privacy inherent in digital communications and discuss potential risks, such as the possibility of unauthorized access to their data. Informed consent documents should explicitly address privacy concerns and explain measures taken to safeguard confidentiality.

Technology Competence and Reliability

Online counseling requires therapists to have adequate technology competence to effectively use digital platforms for therapeutic purposes. This competence includes knowledge of the specific platforms, familiarity with technical troubleshooting, and awareness of any potential limitations or risks associated with using technology in therapy.

Therapists should consider their technological skills and be cautious about using unfamiliar or unreliable platforms that may compromise the quality of the therapeutic relationship. They should also be prepared to address any technical issues that may arise during sessions, ensuring prompt and effective resolution to minimize client distress.

Informed Consent and Boundaries

Obtaining informed consent is a crucial ethical requirement in counseling, regardless of the mode of delivery. However, in online counseling, therapists must explicitly cover aspects unique to the digital context. This includes informed consent for the use of electronic platforms, potential limitations of online therapy, and instructions for maintaining privacy and confidentiality.

Therapists should also establish clear boundaries related to communication methods, response times, emergency contacts, and alternative modes of contact in case of technological disruptions. Setting these boundaries helps maintain a structured therapeutic relationship and ensures that clients understand the expectations and limitations of online counseling.

Emergencies and Crisis Intervention

Crisis intervention is an essential component of mental health care, and online counseling should be prepared to address emergencies that may arise during sessions. It is crucial for therapists to establish emergency protocols and plans for managing critical situations, such as suicidal ideation, self-harm, or the disclosure of abuse.

Therapists should inform clients about the limitations of online counseling in handling emergencies and provide alternative crisis resources such as helpline numbers or local emergency contacts. They should also have contingency plans in place, such as having access to a supervisor or colleague who can provide assistance if needed.

Ethical Considerations in Online Assessment

Assessment is an integral part of mental health care, and online counseling may rely on various assessment tools administered remotely. Ethical considerations in online assessment include ensuring the validity and reliability of assessment measures and safeguarding clients' privacy and confidentiality during the process.

Therapists should be knowledgeable about the appropriateness and limitations of different online assessment tools and carefully select measures that align with the client's needs and therapeutic goals. Informed consent should cover the purpose of assessment, the specific tools to be used, and how the results will be interpreted and shared with the client.

Cultural Competence and Diversity

Cultural competence is essential in online counseling to ensure that therapists can provide appropriate, sensitive, and inclusive care to clients from diverse backgrounds. Therapists should be aware of their own cultural biases, assumptions, and limitations when working with clients online.

Therapists need to actively engage in ongoing education and self-reflection to enhance their cultural competence. This includes understanding the cultural nuances of the clients they work with, adapting therapeutic interventions to different cultural contexts, and addressing potential power imbalances in the online therapeutic relationship.

Continuity of Care and Referrals

Online counseling may not be suitable for all clients or all therapeutic issues. Therapists should be aware of the limitations of online counseling and be prepared to make appropriate referrals when necessary. This may involve referring clients to local resources for face-to-face therapy or to specialized professionals for specific issues outside the scope of their expertise.

Therapists should establish clear guidelines for assessing client suitability for online counseling and regularly evaluate the appropriateness and effectiveness of the online modality. They should also maintain open and honest communication with clients about the potential need for transitioning to in-person therapy or seeking additional support from other professionals.

In conclusion, online counseling offers convenient access to mental health care, but it also poses distinct ethical challenges. Therapists must pay close attention to issues such as confidentiality, technology competence, informed consent, crisis intervention, cultural competence, and continuity of care. By navigating these ethical issues thoughtfully and responsibly, therapists can provide high-quality online counseling that supports clients' well-being and meets professional standards.

Ethical Issues in the Global Context of Health and Wellness

Global Health Inequalities and Global Justice

Health Disparities Between Countries

Health disparities between countries are a significant concern in global health and wellness. These disparities refer to the unequal distribution of health outcomes, healthcare access, and resources among different countries. While some countries have made significant progress in improving health and wellness outcomes, many others continue to face significant challenges.

Understanding Health Disparities

Health disparities are influenced by various factors, including social, economic, cultural, and political determinants of health. These determinants can contribute to differences in health status, access to healthcare services, and overall well-being between countries. Some key factors contributing to health disparities between countries include:

+ **Economic Factors:** Countries with lower income levels and higher levels of poverty often experience greater health disparities. Limited economic resources can result in reduced access to healthcare facilities, medical equipment, medications, and trained healthcare professionals.

+ **Infrastructure and Resources:** Disparities in healthcare infrastructure and resources also contribute to health disparities between countries. Some countries may lack sufficient healthcare facilities, medical technologies, adequate sanitation systems, and clean water supply, which can significantly affect health outcomes.

+ **Education and Awareness:** Differences in educational attainment and health literacy among populations can impact health disparities. Countries with low levels of education and health awareness may struggle to implement effective health promotion strategies, disease prevention programs, and access to accurate health information.

+ **Political Factors:** Political stability, governance, and policies play a crucial role in addressing health disparities. Countries with unstable political systems may struggle to allocate resources effectively, leading to unequal access to healthcare services and disparities in health outcomes.

+ **Social and Cultural Factors:** Social norms, cultural beliefs, and discrimination can also contribute to health disparities. Marginalized populations, such as ethnic and racial minorities, indigenous communities, and migrants, may experience barriers to healthcare access, culturally insensitive care, and unequal treatment.

Addressing Health Disparities

Reducing health disparities between countries requires a comprehensive approach that addresses the underlying factors contributing to these disparities. Some strategies that can help address health disparities include:

+ **Promoting Economic Development:** Improving the economic status of countries can contribute to better access to healthcare resources and services. International efforts, such as foreign aid, investment in healthcare infrastructure, and poverty reduction initiatives, can help alleviate health disparities.

+ **Strengthening Healthcare Systems:** Building robust healthcare systems is essential for addressing health disparities. This includes expanding healthcare infrastructure, ensuring an adequate healthcare workforce, improving healthcare financing mechanisms, and implementing effective health policies and regulations.

+ **Increasing Access to Healthcare:** Enhancing access to quality healthcare services is crucial in reducing health disparities. This can be achieved through the establishment of primary healthcare centers, the provision of essential medications and medical technologies, and the promotion of universal health coverage.

+ **Improving Health Education and Awareness:** Educating populations about health promotion, disease prevention, and the importance of seeking timely healthcare can help reduce health disparities. This includes targeted health education initiatives, culturally sensitive health campaigns, and the dissemination of accurate health information.

+ **Addressing Social and Cultural Determinants:** Tackling social and cultural determinants of health is essential. This involves combating discrimination, promoting social inclusion, recognizing and addressing disparities faced by marginalized populations, and ensuring culturally competent healthcare services.

+ **Fostering Global Collaboration:** International cooperation is critical in addressing health disparities. Collaboration between countries, organizations, and stakeholders can facilitate the sharing of resources, knowledge, and best practices, as well as the development of global health policies aimed at reducing health disparities.

Case Study: Access to Essential Medicines

Access to essential medicines is a critical aspect of healthcare and is often affected by health disparities between countries. Many low- and middle-income countries struggle to provide necessary medications to their populations due to various challenges, including high costs, limited infrastructure, and inadequate healthcare systems. This lack of access to essential medicines can have severe consequences on health outcomes.

For instance, in countries with high prevalence rates of infectious diseases like HIV/AIDS, access to antiretroviral medications can be a matter of life and death. However, the cost of these medications often makes them inaccessible for many individuals in low-income countries. This creates a significant disparity in HIV/AIDS treatment outcomes between countries with well-established healthcare systems and those with limited resources.

Addressing these disparities requires a multi-faceted approach. It involves negotiating fair pricing agreements with pharmaceutical companies, strengthening local healthcare systems to ensure efficient drug distribution, and investing in research and development of affordable generic medications. Additionally, international collaborations and partnerships can contribute to improving access to essential medicines by providing financial and technical support to countries in need.

Conclusion

Health disparities between countries pose significant challenges to achieving global health and wellness. These disparities are influenced by a complex interplay of social, economic, cultural, and political factors. Addressing health disparities requires comprehensive strategies that encompass economic development, strengthening healthcare systems, increasing access to healthcare, improving health education, addressing social and cultural determinants, and fostering global collaboration. By working together towards reducing health disparities, we can strive for a more equitable and just health and wellness future for all.

Access to Essential Medicines and Global Justice

Access to essential medicines is a fundamental aspect of global health and wellness. However, there are significant challenges in ensuring equitable access to these life-saving drugs, particularly in low-income countries. This section explores the ethical issues surrounding access to essential medicines and the concept of global justice.

The Importance of Access to Essential Medicines

Access to essential medicines is crucial for the promotion of health and the treatment of diseases. Essential medicines are those drugs that are deemed necessary to meet the health needs of the population, based on their prevalence and effectiveness in treating specific diseases. These medicines are often included in the World Health Organization's (WHO) Model List of Essential Medicines.

Ensuring access to essential medicines is essential for several reasons. Firstly, it is a matter of human rights. Article 12 of the International Covenant on Economic, Social and Cultural Rights recognizes "the right of everyone to the enjoyment of the highest attainable standard of physical and mental health." Access to essential medicines is an integral component of the realization of this right.

Secondly, access to essential medicines is critical for achieving global health goals, such as the Sustainable Development Goals (SDGs). The SDGs aim to ensure healthy lives and promote well-being for all at all ages. Access to essential medicines plays a vital role in reducing maternal and child mortality, combating infectious diseases, and improving overall health outcomes.

Challenges to Access

Despite the importance of access to essential medicines, numerous challenges hinder its realization, particularly in low-income countries. These challenges include:

+ **High Costs:** Many essential medicines are produced by pharmaceutical companies that hold patents, making them costly and inaccessible to resource-constrained countries.

+ **Intellectual Property Rights:** The TRIPS Agreement (Trade-Related Aspects of Intellectual Property Rights) places restrictions on the production and distribution of generic medicines, limiting competition and driving up prices.

+ **Lack of Infrastructure:** Inadequate healthcare infrastructure and limited healthcare resources in low-income countries pose challenges for the procurement, storage, and distribution of essential medicines.

+ **Supply Chain Issues:** Weak supply chains and logistical challenges can lead to stockouts and shortages of essential medicines, undermining access and treatment continuity.

+ **Limited Research and Development:** There is a lack of investment in research and development for medicines that primarily affect low-income countries, such as neglected tropical diseases.

Ethical Considerations

Addressing the challenges to access to essential medicines requires a comprehensive approach rooted in ethical considerations. The principles of justice, equity, and solidarity are particularly relevant in this context.

Justice: The principle of justice calls for fair and equitable distribution of resources. In the case of access to essential medicines, justice entails ensuring that everyone, regardless of their socio-economic status or geographic location, has access to the medicines they need to live a healthy life. This requires addressing the structural inequalities that hinder access, such as high drug prices and intellectual property rights.

Equity: Equity recognizes that different individuals and communities have different needs and requires allocating resources to achieve fairness. In the context of access to essential medicines, equity demands prioritizing the needs of the most vulnerable populations, such as those in low-income countries or marginalized communities within wealthier nations.

Solidarity: Solidarity emphasizes collective responsibility and cooperation in addressing health challenges. It requires a global commitment to ensuring access to essential medicines as a shared responsibility. This includes fostering partnerships between governments, non-governmental organizations, private companies, and international bodies to develop sustainable solutions and overcome barriers to access.

Promoting Access to Essential Medicines

Several strategies can help promote access to essential medicines and advance global justice:

+ **Reducing Drug Prices:** Governments can negotiate with pharmaceutical companies to lower prices or promote the use of generic medicines. This can be facilitated through the use of compulsory licensing, which allows the production of generic versions of patented medicines.

+ **Strengthening Health Systems:** Investing in healthcare infrastructure, supply chain management, and workforce capacity can enhance the availability and accessibility of essential medicines.

+ **Research and Development (R&D) Reform:** Encouraging R&D that focuses on diseases affecting low-income countries and providing incentives for the development of affordable medicines can address the gaps in treatment options.

+ **International Cooperation:** Strengthening international collaborations and partnerships can facilitate technology transfer, capacity building, and resource sharing to improve access to essential medicines.

+ **Humanitarian Licenses:** Allowing the production and distribution of essential medicines without patent barriers during public health emergencies can ensure timely access to life-saving treatments.

Case Study: Access to HIV/AIDS Medications

The issue of access to essential medicines gained significant attention during the HIV/AIDS epidemic. In the 1990s, the high cost of antiretroviral drugs made them inaccessible to millions of people, predominantly in low-income countries.

However, through advocacy efforts and international collaboration, access to HIV/AIDS medications has significantly improved. Initiatives like the President's Emergency Plan for AIDS Relief (PEPFAR) and the Global Fund to Fight AIDS, Tuberculosis, and Malaria have played a pivotal role in increasing access to life-saving medications.

Conclusion

Access to essential medicines is not only a matter of health and well-being but also a question of ethics and justice. Addressing the challenges to access requires a multi-dimensional approach, involving legal, political, and social reforms. By promoting equity, justice, and solidarity, it is possible to create a world where essential medicines are accessible to all, regardless of their geographic location or socio-economic status.

Global Health Governance and Ethics

Global health governance refers to the collective efforts and mechanisms put in place to address health issues on a global scale, with the aim of promoting health equity and improving health outcomes for all individuals, regardless of their socioeconomic status or geographic location. In this section, we will explore the ethical considerations that arise in the context of global health governance and discuss the challenges and potential solutions related to achieving global health equity.

The Importance of Global Health Governance

Global health governance plays a vital role in addressing health challenges that extend beyond national boundaries. It recognizes that health issues such as infectious diseases, environmental hazards, and inadequate healthcare systems do not respect geopolitical boundaries and require collaborative efforts to address effectively. Global health governance seeks to ensure that all individuals, regardless of their nationality, have access to essential healthcare services and can enjoy the highest attainable standard of health.

Ethical principles such as justice, solidarity, and human rights form the foundation of global health governance. These principles emphasize the importance of addressing health inequities and working towards fair distribution of health resources globally. They also recognize the right to health as a fundamental human right, which requires governments, international organizations, and other stakeholders to take responsibility for promoting and protecting health on a global scale.

Challenges in Global Health Governance

While the need for global health governance is clear, there are several challenges that hinder its effectiveness. These challenges include:

1. **Power imbalances and inequalities:** Power imbalances between high-income and low-income countries, as well as within countries themselves, often result in disproportionate allocation of resources and unequal distribution of health benefits. This exacerbates health disparities and leaves marginalized populations at a higher risk of poor health outcomes.

2. **Lack of coordination:** Global health governance involves the collaboration of numerous stakeholders, including governments, international organizations, civil society, and the private sector. However, the lack of coordination and coherence among these actors hampers effective response to global health challenges. This can lead to duplication of efforts, inefficiencies, and fragmented healthcare systems.

3. **Limited resources and funding:** Insufficient funding and resources pose a significant challenge to global health governance. Many low-income countries struggle to allocate adequate resources to their healthcare systems, making it difficult to address health issues effectively. Additionally, funding for global health initiatives often faces competing priorities and donor fatigue, limiting the overall impact that can be achieved.

4. **Political barriers:** Political factors, including conflicts, geopolitical rivalries, and differing national interests, can pose obstacles to effective global health governance. These barriers can impede cooperation, hinder resource mobilization, and delay the implementation of necessary health interventions.

5. **Inequitable access to medicines and technologies:** Limited access to essential medicines and health technologies is a critical challenge in global health governance. Intellectual property rights, high costs, and lack of infrastructure contribute to inequitable access to life-saving treatments and technologies, particularly for individuals in low-income countries.

Ethical Principles in Global Health Governance

To address these challenges and promote ethical global health governance, several key principles need to be considered:

1. **Justice and equity:** Justice requires that health resources and interventions be distributed fairly, ensuring that the most vulnerable populations have access to needed services. Equity emphasizes the need to address social and structural determinants of health that contribute to health disparities globally. Efforts should focus on reducing inequalities in health outcomes and opportunities for health.

2. **Solidarity:** Solidarity recognizes that health is a shared responsibility and that global health challenges can only be effectively addressed through collective action. International cooperation, resource sharing, and knowledge exchange are essential components of solidarity in global health governance.

3. **Participation and inclusivity:** Recognizing the diverse perspectives and experiences of various stakeholders, including marginalized populations, is crucial for ethical global health governance. Meaningful engagement and participation in decision-making processes ensure that the voices of all individuals, especially those most affected by health issues, are heard and valued.

4. **Accountability and transparency:** Ethical global health governance requires clear mechanisms of accountability and transparency. Governments, organizations, and other stakeholders should be accountable for their actions and decisions, and transparent in their allocation of resources, decision-making processes, and reporting of outcomes. This fosters trust among stakeholders and enhances the effectiveness of global health initiatives.

5. **Human rights:** Global health governance should be guided by a human rights approach, recognizing that health is a fundamental human right. This necessitates the protection of individuals' rights to health, non-discrimination, privacy, and access to information. Human rights-based approaches empower individuals to claim their rights and hold duty bearers accountable for fulfilling their obligations.

Examples and Solutions

To illustrate the challenges and potential solutions in global health governance, let's consider two current examples: access to COVID-19 vaccines and the global response to HIV/AIDS.

Access to COVID-19 Vaccines: The COVID-19 pandemic has highlighted the importance of equitable access to vaccines. Ensuring global vaccine equity is essential to control the spread of the virus and minimize its devastating impact on health systems and economies worldwide. However, challenges such as limited vaccine supply, vaccine nationalism, and price disparities hinder equitable access. To address these challenges, countries and international organizations can:

+ Collaborate to increase production and distribution of vaccines, including technology transfer and intellectual property sharing, to overcome supply shortages.

+ Establish funding mechanisms to support the procurement of vaccines for low-income countries.

+ Advocate for fair pricing and the removal of trade barriers to make vaccines more affordable and accessible globally.

+ Promote vaccine equity through global initiatives, such as COVAX, which aims to ensure fair and equitable access to COVID-19 vaccines.

Global Response to HIV/AIDS: The global response to HIV/AIDS provides another example of the challenges and solutions in global health governance. The initial response to the HIV/AIDS epidemic highlighted the importance of human rights, community mobilization, and access to treatment. International cooperation, advocacy, and funding have been instrumental in addressing the HIV/AIDS crisis. However, challenges such as stigma and discrimination, limited resources, and sustaining long-term interventions persist. To address these challenges, stakeholders can:

+ Promote education and awareness to reduce HIV/AIDS-related stigma and discrimination, ensuring that all individuals, regardless of their HIV status, are treated with dignity and respect.

+ Increase funding for prevention and treatment programs, including the Global Fund to Fight AIDS, Tuberculosis, and Malaria.

+ Strengthen health systems in low-income countries to ensure sustainable and comprehensive HIV/AIDS care, including the availability of antiretroviral therapy.

+ Support community-based organizations and initiatives that play a crucial role in HIV/AIDS prevention, care, and support services.

Conclusion

Global health governance and ethics are intertwined, promoting health equity and addressing health challenges on a global scale. Ethical principles such as justice, solidarity, participation, and accountability guide the efforts towards a more equitable, just, and sustainable global health future. Overcoming challenges in global health governance requires collective action, political will, and international cooperation. By prioritizing ethical considerations and addressing health inequities, we can strive towards a world where everyone has the opportunity to enjoy good health and well-being.

Infectious Diseases and Global Health

Ethical Considerations in Disease Prevention and Control

Disease prevention and control are vital components of public health efforts aimed at reducing the burden of illness and promoting the well-being of communities. However, the implementation of disease prevention and control strategies raises ethical considerations that need to be carefully addressed. These ethical considerations revolve around issues such as balancing the rights of individuals with the need to protect public health, ensuring equitable access to prevention measures, addressing potential conflicts of interest, and respecting cultural and societal values. In this section, we will explore some of these ethical considerations in the context of disease prevention and control.

Balancing Individual Rights and Public Health

One of the fundamental ethical tensions in disease prevention and control is the need to balance individual rights and autonomy with the broader goal of protecting public health. On the one hand, individuals have the right to make decisions about their own health, including whether or not to participate in disease prevention measures. This principle of individual autonomy is central to medical ethics and respects individuals' right to make informed choices about their own bodies.

However, in the context of infectious diseases, individual choices can have significant implications for public health. For example, individuals who choose not to be vaccinated against a highly contagious and potentially deadly disease like measles can put vulnerable populations at risk. This tension between individual autonomy and public health is often resolved through the implementation of policies and regulations that enforce vaccination or other preventive measures.

Example: In 2019, a measles outbreak occurred in several communities in the United States. Many of the affected individuals were unvaccinated due to personal or religious beliefs. In response to the outbreak, several states passed laws that eliminated non-medical exemptions for school vaccination requirements. These laws aimed to protect public health by ensuring high vaccination rates among school-age children.

Equitable Access to Prevention Measures

In the context of disease prevention and control, it is essential to ensure equitable access to preventive measures to avoid exacerbating existing health disparities. Preventive measures such as vaccinations, screenings, and health education campaigns are critical for reducing the burden of disease and promoting population health. However, certain populations, such as those living in low-income communities or marginalized groups, may face barriers to accessing these preventive measures.

Ethical considerations arise when some individuals or communities are disproportionately burdened by disease due to limited access to preventive measures. In such cases, efforts should be made to address the underlying social determinants of health that contribute to these disparities and ensure that preventive measures are accessible to all populations, regardless of socioeconomic status or other demographic factors.

Example: In response to the COVID-19 pandemic, many countries implemented mass vaccination campaigns to curb the spread of the virus. However, inequitable access to vaccines became a significant ethical concern. Wealthier nations secured large quantities of vaccines, while less affluent countries struggled to obtain an adequate supply. This highlighted the need for global collaboration and resource allocation to ensure equitable access to vaccines for all populations.

Conflicts of Interest in Disease Prevention and Control

Conflicts of interest can arise in the field of disease prevention and control, particularly when public health decisions are influenced by commercial or political interests. Conflicts of interest occur when individuals or organizations have competing interests that could potentially compromise the integrity of public health decision-making processes.

For example, pharmaceutical companies may sponsor research studies or public health campaigns related to disease prevention and control. While this can provide valuable resources and expertise, it also raises concerns about the potential influence of industry on public health policies and recommendations. Transparent and robust mechanisms to identify and manage conflicts of interest are essential to ensure that public health decisions are made based on scientific evidence and the best interests of the population.

Example: The tobacco industry has a long history of engaging in activities that undermine public health efforts to prevent smoking-related diseases. Through advertising, lobbying, and funding of research, tobacco companies have sought to influence public opinion and policy-making in ways that prioritize their profits over public health. Identifying and addressing these conflicts of interest is crucial to maintaining the integrity of disease prevention and control efforts.

Respecting Cultural and Societal Values

Disease prevention and control strategies should be sensitive to the cultural, societal, and religious values of the populations they aim to serve. Different communities may have varying beliefs, practices, and norms that influence their acceptance and participation in preventive measures. Respecting these values is key to building trust, fostering engagement, and ensuring the effectiveness of prevention efforts.

Efforts to respect cultural and societal values should involve meaningful community engagement and collaboration, and should take into account the diverse perspectives and experiences of the affected populations. Cultural competence, which entails understanding and incorporating cultural diversity into health practices, is essential in designing and implementing disease prevention and control strategies that are respectful and relevant to the communities they serve.

Example: In some cultures, traditional healing practices are deeply rooted and play a significant role in disease prevention and control. For instance, in some indigenous communities, cultural ceremonies and practices may be central to their overall well-being and prevention efforts. Understanding and respecting these traditions can help in developing culturally appropriate disease prevention strategies that are accepted and trusted by the community.

Resource Allocation and Prioritization

Disease prevention and control efforts often require resource allocation and prioritization, particularly in resource-limited settings. The challenge is to balance the allocation of limited resources, such as funding, personnel, and infrastructure, while maximizing the overall impact on population health.

Ethical considerations arise when decisions need to be made about how resources are allocated and prioritized. Principles of distributive justice, which aim to ensure fairness in resource distribution, should guide these decisions. Health equity, efficiency, and the potential for improving health outcomes should be considered when prioritizing the allocation of resources for disease prevention and control.

Example: In the context of a cholera outbreak, health authorities may need to decide how to allocate limited supplies of oral rehydration solution (ORS) between different affected communities. Ethical considerations would involve assessing factors such as the severity of the outbreak, the vulnerability of the affected populations, and the potential impact of providing ORS on reducing mortality and morbidity.

In conclusion, ethical considerations play a crucial role in disease prevention and control efforts. Balancing individual rights with the need to protect public health, ensuring equitable access to preventive measures, addressing conflicts of interest, respecting cultural and societal values, and making decisions about resource allocation and prioritization are vital ethical considerations that need to be thoughtfully addressed. By taking these ethical considerations into account, we can strive to implement disease prevention and control strategies that are effective, equitable, and respectful of individuals and communities.

Vaccine Ethics and Mandatory Vaccination

Vaccination is one of the most effective ways to prevent the spread of infectious diseases and protect public health. However, the issue of vaccine ethics and mandatory vaccination has sparked debates and controversies in recent years. In this section, we will explore the ethical considerations surrounding vaccine policies, the concept of mandatory vaccination, and the balance between individual rights and public health.

The Importance of Vaccination

Before delving into the ethical aspects of mandatory vaccination, it is crucial to understand the importance of vaccination in promoting individual and public health. Vaccines have been instrumental in preventing the spread of diseases such as measles, polio, and influenza. They not only protect the individuals who receive them but also contribute to herd immunity, which shields vulnerable populations who cannot receive vaccines due to medical reasons.

Vaccination also plays a significant role in reducing the burden on healthcare systems and minimizing healthcare costs. By preventing the occurrence and transmission of diseases, vaccines help prevent hospitalizations, complications, and long-term disability associated with infectious diseases.

Ethical Principles in Vaccine Policies

When formulating vaccine policies, several ethical principles come into play. These principles guide decision-making and ensure that vaccination programs are effective, equitable, and respectful of individuals' rights. Some key ethical principles in vaccine policies include:

1. **Beneficence:** The principle of beneficence emphasizes the obligation to promote the well-being of individuals and the greater population. Vaccination is considered a beneficent act as it prevents harm and promotes health.

2. **Autonomy:** The principle of autonomy recognizes individuals' rights to make informed decisions about their healthcare. It acknowledges that individuals have the freedom to accept or refuse vaccines, based on their own values and beliefs.

3. **Justice:** The principle of justice requires fair and equitable distribution of healthcare resources. This principle calls for equal access to vaccines and protection against vaccine-preventable diseases.

Mandatory Vaccination

Mandatory vaccination refers to policies that require individuals to receive certain vaccines, usually with exemptions for medical and religious reasons. The rationale behind mandatory vaccination is to protect both individuals and the population as a whole and uphold the principle of justice by ensuring equal access to healthcare.

Proponents of mandatory vaccination argue that it is necessary to achieve herd immunity, especially for individuals who cannot receive vaccines due to medical conditions. They believe that the public health benefits outweigh the individual's right to refuse vaccination.

However, opponents of mandatory vaccination raise concerns about individual autonomy and freedom of choice. They argue that individuals should have the right to make decisions about their own bodies and medical treatments, even if it poses potential risks to public health. Some also express concerns about vaccine safety and side effects.

Balancing Individual Rights and Public Health

The question of balancing individual rights and public health is at the core of the ethical considerations surrounding mandatory vaccination. Striking a balance between individual autonomy and the greater good of public health is a challenging task for policymakers and healthcare professionals.

Ethically, it is important to respect and promote individual autonomy while safeguarding public health. Providing accurate and accessible information about vaccines, addressing concerns, and addressing vaccine hesitancy through effective communication strategies can empower individuals to make informed decisions about vaccination.

It is also essential to ensure that vaccine policies do not disproportionately burden vulnerable populations or perpetuate existing health disparities. Efforts should be made to address barriers to vaccination, including access issues, cultural and religious beliefs, and distrust in healthcare systems.

Real-World Examples

To illustrate the ethical considerations in vaccine ethics and mandatory vaccination, let's consider two real-world examples:

Example 1: Measles Outbreak

During a measles outbreak in a community, health authorities implement a mandatory vaccination policy to curb the spread of the disease. Individuals who refuse vaccination without a valid medical or religious exemption are subject to penalties, such as exclusion from schools or public spaces.

In this scenario, the ethical principle of beneficence guides the decision to implement mandatory vaccination as a measure to prevent harm and protect public health. However, the principle of autonomy may be challenged, as individuals are obliged to comply with the mandatory policy.

Example 2: COVID-19 Vaccination

In response to the COVID-19 pandemic, governments around the world launch mass vaccination campaigns. While the majority of individuals willingly get vaccinated, a small percentage of the population expresses concerns about the safety and efficacy of the vaccines.

The ethical challenge here lies in balancing the individual's right to autonomy and freedom of choice with the need to achieve herd immunity and control the pandemic. Effective communication, addressing concerns, and ensuring equitable access to vaccines can help promote vaccination while respecting individual autonomy.

Resources and Further Reading

For further exploration of vaccine ethics and mandatory vaccination, the following resources may be helpful:

1. "Vaccine Ethics: An Introduction" by Jason L. Schwartz

2. "The Ethics of Vaccination" by Alberto Giubilini and Thomas Douglas

3. "Vaccine Ethics and Policy: A Guide for Global Health Communication" by L.W. Ortega, et al.

Exercise

Reflect on your understanding of vaccine ethics and mandatory vaccination by considering the following questions:

1. What ethical principles should guide vaccine policies?

2. How can concerns about individual autonomy be addressed in mandatory vaccination programs?

3. What strategies can be implemented to promote vaccine acceptance and reduce vaccine hesitancy?

Take a moment to jot down your thoughts and explore these questions further to deepen your understanding of the ethical dimensions of vaccine ethics and mandatory vaccination.

In conclusion, the ethical considerations surrounding vaccine ethics and mandatory vaccination revolve around finding a balance between individual rights and public health. While mandatory vaccination policies can promote the greater good, it is crucial to respect individual autonomy, address concerns, and ensure equitable access to vaccines. By upholding ethical principles and promoting effective communication, we can navigate the complex terrain of vaccine ethics and strive for a healthier and more resilient society.

Pandemics and Allocation of Scarce Resources

In times of pandemics, such as the global COVID-19 crisis, healthcare systems often face the challenging task of allocating scarce resources. As the demand for medical services and interventions exceeds the available capacity, difficult decisions need to be made regarding who receives treatment and who does not. These allocation decisions raise profound ethical questions and require careful consideration of principles such as justice, beneficence, and utility.

The Principle of Justice in Resource Allocation

The principle of justice plays a crucial role in guiding the allocation of scarce resources during pandemics. This principle emphasizes fairness and equity in the distribution of healthcare resources. Equality, need, and social worth are all factors that can be considered in ensuring a just allocation process.

One approach to resource allocation is based on egalitarian principles of equality and fairness. This approach suggests that resources should be distributed equally among all individuals, regardless of factors such as age or underlying health conditions. It argues that everyone should have an equal chance of receiving life-saving treatments. However, this approach may not take into account the varying degrees of need and potential benefit that different individuals may have.

Another approach is a needs-based allocation system, which prioritizes those in the greatest need. This can be determined by assessing the severity of illness, the likelihood of survival, or the urgency of treatment. By prioritizing those with the highest need, this approach aims to save as many lives as possible. However, it may result in excluding individuals who could benefit from treatment but are deemed to have a lower immediate need.

A third approach involves considering the social worth or contribution of individuals to society. This approach takes into account the role that individuals play in society, such as frontline healthcare workers or essential service providers. It argues that those who contribute more should receive priority in resource allocation. However, this approach raises concerns of discrimination and marginalization, as it may disadvantage groups who are already disadvantaged or vulnerable.

Finding a balance between these different principles is necessary to ensure a just allocation process. It is important to consider the contextual factors and the specific needs of the population being served. Ethical guidelines and committees can play a crucial role in providing guidance and ensuring fairness in resource allocation decisions.

Ethical Challenges in Resource Allocation

The allocation of scarce resources during pandemics presents several ethical challenges. These challenges include difficult decision-making, transparency and accountability, and the potential for bias and discrimination.

One of the most significant challenges is the ethical dilemma faced by healthcare providers and policymakers when making allocation decisions. These decisions often involve choosing between individuals who are equally in need of medical interventions, but resources are insufficient to treat all. This can lead to immense ethical and emotional burdens on healthcare providers, as they must prioritize some patients while denying treatment to others.

Transparency and accountability are essential in resource allocation. Clear and explicit guidelines should be developed and communicated to the public, explaining the allocation process and the ethical reasoning behind it. Open dialogue and engagement with the public can help build trust and address concerns about fairness and equity.

Another challenge is the potential for bias and discrimination. In resource allocation, there is a risk that decisions may be influenced by implicit biases, such as ageism or systemic inequalities. It is crucial to guard against such biases and ensure that decisions are based on objective criteria and evidence-based guidelines.

Strategies for Ethical Resource Allocation

To address the ethical challenges in resource allocation during pandemics, several strategies can be employed.

Firstly, developing clear and transparent guidelines that reflect ethical principles is crucial. These guidelines should be developed in advance, with input from experts in various fields, including ethics, public health, and medicine. The guidelines should provide a structured framework for decision-making, taking into account factors such as prognosis, potential benefit, and social worth.

Secondly, inclusive and diverse decision-making bodies should be established to ensure a broader representation of perspectives. These bodies can include healthcare professionals, ethicists, community representatives, and vulnerable populations. Involving diverse voices can help mitigate bias and discrimination and ensure a fair and just allocation process.

Thirdly, ongoing ethical reflection and review of allocation decisions are necessary. Regular evaluation and feedback mechanisms can assist in identifying and addressing any potential biases or flaws in the allocation process. This continuous improvement process helps maintain transparency and integrity in resource allocation.

Lastly, public engagement and education are vital in facilitating understanding and acceptance of the allocation process. Clear communication about the rationale behind resource allocation decisions can help alleviate public anxieties and foster a sense of solidarity and shared responsibility.

Case Study: Allocation of Ventilators during COVID-19

The allocation of ventilators during the COVID-19 pandemic provides a real-world example of the ethical challenges and considerations involved in resource allocation.

In the early stages of the pandemic, many healthcare systems faced a shortage of ventilators. As a life-saving intervention for severe respiratory distress, decisions needed to be made about which patients would receive access to these limited resources.

Ethical guidelines were developed to assist healthcare providers in making allocation decisions. These guidelines took into account factors such as the severity of illness, the likelihood of survival, and the potential for long-term health outcomes. Some guidelines also considered the principle of social worth, prioritizing healthcare workers who were on the frontline of the pandemic response.

While these guidelines provided a framework, the decision-making process was still challenging. In some cases, it involved difficult choices between patients with similar prognoses. Healthcare providers, burdened with the responsibility of making life-or-death decisions, faced significant moral distress.

The allocation of ventilators also highlighted underlying systemic inequalities and disparities in access to healthcare. Marginalized communities, including racial and ethnic minorities, were disproportionately affected by the virus and had increased difficulty accessing vital resources.

Overall, the allocation of ventilators during the COVID-19 pandemic demonstrated the need for clear ethical guidelines, transparency, ongoing review, and public engagement. It also underscored the importance of addressing structural inequalities to ensure a just and equitable allocation process.

Summary

The allocation of scarce resources during pandemics raises complex ethical questions and challenges. The principle of justice, grounded in fairness and equity, plays a vital role in guiding resource allocation decisions. Approaches based on equality, need, and social worth can help inform allocation processes.

Ethical challenges in resource allocation include the difficult decision-making process, transparency and accountability, and the potential for bias and discrimination. Strategies such as clear guidelines, inclusive decision-making bodies, ongoing ethical reflection, and public engagement can help address these challenges.

The allocation of ventilators during the COVID-19 pandemic serves as a real-world case study. It highlights the importance of ethical guidelines, the burdens faced by healthcare providers, and the need to address systemic inequalities.

By navigating the ethical dilemmas inherent in resource allocation, we can strive to ensure a fair and just allocation of scarce resources during pandemics, ultimately working towards a healthier and more equitable world.

Ethics of Medical Tourism

Risks and Benefits of Medical Tourism

Medical tourism, also known as health tourism or medical travel, refers to the practice of traveling to a foreign country to receive medical treatment or services. It has become increasingly popular in recent years, as people seek high-quality and cost-effective healthcare options outside their home countries. However, along with its benefits, medical tourism also presents several risks and challenges.

Benefits of Medical Tourism

1. Cost Savings: One of the primary reasons why individuals opt for medical tourism is the potential for cost savings. Many countries offer medical treatments at significantly lower prices compared to Western countries, making it an attractive option for those seeking affordable healthcare.

2. Access to High-Quality Care: Medical tourism allows patients to access medical treatments and procedures that may not be available or are limited in their home countries.   This includes specialized medical expertise, advanced technologies, and cutting-edge procedures.

3. Reduced Waiting Times: In countries with publicly funded healthcare systems, long waiting times for non-emergency medical procedures are common. Medical tourism offers the benefit of reduced waiting times, allowing patients to receive timely treatment and avoid prolonged discomfort or worsening of their health conditions.

4. Combination of Treatment and Vacation: Medical tourism often provides an opportunity for patients to combine their medical treatment with a vacation. This allows individuals to relax and recuperate in a new environment, potentially enhancing their overall well-being and recovery.

5. Cultural and Language Considerations: Some patients may choose medical tourism to seek healthcare services that align with their cultural, religious, or language preferences.   By selecting a destination that shares their cultural background or speaks their language, patients may feel more comfortable and better understood during their medical journey.

Risks of Medical Tourism

1.   Quality and Safety Concerns: While medical tourism offers access to high-quality care in certain destinations, the quality and safety standards can vary significantly between countries and healthcare providers.   It is important for patients to thoroughly research and evaluate the qualifications, expertise, and reputation of healthcare facilities and providers to ensure they receive reliable and safe care.

2.   Communication and Language Barriers: Language barriers can pose challenges in medical tourism, as effective communication between patients and healthcare providers is crucial for accurate diagnoses, treatment planning, and post-treatment care.   Miscommunication can lead to misunderstandings and inadequate care.

3.   Legal and Regulatory Differences: Legal and regulatory standards for medical practice vary across countries. Patients considering medical tourism should familiarize themselves with the local laws and regulations governing medical treatments, including issues related to liability, malpractice, and patient rights.

4. Follow-up Care and Continuity: After undergoing medical treatment abroad, patients may face difficulties with follow-up care and continuity upon returning to their home countries.   Coordination between healthcare providers from different

countries can be challenging, and patients may face difficulties in accessing necessary follow-up consultations and support.

5. Travel Risks and Complications: Medical tourism involves travel to foreign countries, which carries inherent risks such as transportation issues, cultural adjustments, and exposure to unfamiliar environments. Patients need to consider these factors and ensure they are physically fit to travel and undergo medical procedures in a foreign setting.

Addressing the Risks

To minimize the risks associated with medical tourism, individuals should consider the following strategies:

1. Thorough Research: Patients should conduct comprehensive research on potential healthcare destinations, including evaluating the qualifications and credentials of healthcare providers and facilities. Reviewing patient testimonials and seeking recommendations from trusted sources can also provide valuable insights.

2. Consultation with Home Healthcare Providers: Before embarking on medical tourism, patients should consult with their home healthcare providers. This will ensure that their medical needs are adequately assessed and that they receive appropriate guidance regarding treatment options and potential risks.

3. Clear Communication: Patients should have clear and open communication with their healthcare providers, both at home and abroad. This includes discussing their medical history, current health status, and treatment preferences. Patients must ensure they fully understand the proposed treatment plan, associated risks, and expected outcomes.

4. Access to Medical Records: Patients should obtain copies of their medical records, diagnostic test results, and any relevant documentation before traveling for medical treatment. This will facilitate continuity of care and allow healthcare providers abroad to have a comprehensive understanding of the patient's medical history.

5. Travel Insurance and Legal Protection: Patients should consider obtaining comprehensive travel insurance that covers medical emergencies, complications, and any potential legal issues that may arise during their medical journey. Understanding their rights and legal recourse in case of medical malpractice or other adverse events is essential.

6. Post-Treatment Care Plan: Patients should discuss and establish a clear post-treatment care plan with their healthcare providers before returning to their

home countries. This should include provisions for follow-up consultations, necessary medications, and any required rehabilitation or therapy.

7. Transparent Pricing and Billing: Patients should seek transparency in pricing and billing processes from healthcare providers abroad. It is important to understand the full cost of the treatment, including any additional charges or potential hidden costs, to avoid unexpected financial burdens.

Overall, medical tourism can offer significant benefits in terms of cost savings, access to specialized care, and reduced waiting times. However, it is crucial for patients to consider and address the associated risks and challenges through careful planning, research, and open communication with healthcare providers. By taking these precautions, individuals can make informed decisions and increase the likelihood of a positive medical tourism experience.

Equity and Justice in Medical Tourism

Medical tourism, the practice of traveling to another country to receive medical treatment, has become increasingly popular in recent years. People seek medical tourism for a variety of reasons, including cost savings, shorter wait times, access to specialized treatments, and the opportunity for a vacation. However, alongside the benefits, medical tourism raises important ethical concerns regarding equity and justice.

Defining Equity and Justice in Medical Tourism

Equity in medical tourism refers to the fair distribution of healthcare resources and opportunities across different populations. It involves ensuring that all individuals, regardless of their socioeconomic status, have an equal chance to access and benefit from medical tourism. Justice, on the other hand, relates to the ethical principles and societal values governing the allocation and distribution of healthcare resources.

In the context of medical tourism, equity and justice highlight the need to address disparities in access to care between individuals from different socioeconomic backgrounds, both within and between countries. It raises questions about the ethical responsibility of healthcare providers, governments, and the international community to ensure fairness and inclusiveness in medical tourism, while also considering the impact on the host and source countries' healthcare systems.

Challenges to Equity and Justice

One of the main challenges to equity and justice in medical tourism is the unequal access to healthcare resources and opportunities. The high costs associated with travel, accommodation, and medical procedures often exclude individuals from lower socioeconomic backgrounds. This creates a system in which medical tourism primarily benefits affluent individuals, leaving disadvantaged populations without access to the same opportunities for high-quality healthcare.

Moreover, medical tourism can exacerbate existing health disparities between countries. Wealthier countries with advanced medical infrastructure attract patients from countries with limited healthcare resources, leading to brain drain and the further depletion of healthcare personnel and services in the source countries. This creates a cycle of inequity, where the most vulnerable populations in source countries continue to face inadequate healthcare while their wealthier counterparts seek treatment abroad.

Strategies for Promoting Equity and Justice

Addressing the equity and justice concerns in medical tourism requires a multi-faceted approach involving various stakeholders. Here are some strategies to consider:

1. Enhancing local healthcare systems: Rather than solely focusing on attracting medical tourists, governments should prioritize investing in their own healthcare systems. This includes improving infrastructure, training healthcare professionals, and expanding access to high-quality care for all citizens.

2. Reducing costs: Efforts should be made to make healthcare more affordable and accessible within countries. This can involve implementing universal healthcare coverage, price regulation, and subsidies for medical procedures, reducing the need for individuals to seek treatment abroad.

3. Collaboration and knowledge transfer: Developed countries can promote equity and justice in medical tourism by collaborating with source countries to build their healthcare capacity. This can involve sharing knowledge, providing training opportunities, and supporting the development of sustainable healthcare infrastructure.

4. Transparent and ethical practices: Healthcare providers and facilitators involved in medical tourism should adhere to ethical guidelines that prioritize patient welfare, informed consent, and transparency. This includes providing accurate information about treatment options, risks, and costs, and ensuring that patients have realistic expectations.

5. Regulation and oversight: Governments should establish regulatory frameworks to govern medical tourism practices, protecting patients' rights, and ensuring standards of care. This can include accreditation of healthcare facilities, monitoring of outcomes, and addressing legal and ethical issues, such as organ trafficking and exploitation.

Case Study: Equity and Justice in Medical Tourism

To illustrate the challenges and potential solutions related to equity and justice in medical tourism, consider the case of cosmetic surgery tourism. People from Western countries sometimes travel to developing countries for inexpensive cosmetic procedures. While this may provide cost savings for the patients, it raises ethical concerns regarding the exploitation of healthcare workers and the potential negative consequences for local populations.

To promote equity and justice in this scenario, governments and international organizations could collaborate to improve the regulation and oversight of cosmetic surgery tourism. This could involve establishing ethical guidelines for cosmetic surgeons, ensuring appropriate training and qualifications, and implementing safeguards to protect local communities from negative impacts.

Furthermore, efforts could be made to address the underlying reasons why individuals seek cosmetic surgery abroad, such as societal pressures and unrealistic beauty ideals. This can involve public education campaigns promoting body positivity and self-acceptance, as well as regulation of media and advertising industries to prevent the promotion of harmful beauty standards.

Conclusion

Equity and justice are fundamental considerations in the domain of medical tourism. As this industry continues to grow, it is essential to address the disparities and ethical concerns that arise. By adopting strategies that prioritize fairness, collaboration, and the improvement of local healthcare systems, it is possible to mitigate the negative impacts and create a more equitable and just medical tourism landscape.

Regulation and Ethics of Medical Tourism

Medical tourism, the practice of traveling to another country to receive medical care, has gained popularity in recent years due to a variety of factors such as lower costs, shorter waiting times, and access to specialized treatments. However, the rapid growth of this industry has raised concerns about the regulation and ethics surrounding medical tourism. In this section, we will explore the various challenges and considerations in regulating and ensuring ethical practices in the field of medical tourism.

Defining medical tourism

Before delving into the regulatory and ethical aspects, it is important to have a clear understanding of what medical tourism entails. Medical tourism refers to the process of individuals seeking medical treatment or procedures abroad, often with the intention of receiving affordable or high-quality care that may not be readily available in their home country. This could range from elective procedures like cosmetic surgery to complex treatments such as organ transplants or cancer therapies.

Regulatory challenges in medical tourism

Medical tourism presents unique regulatory challenges due to the intersection of multiple legal and cultural systems. The lack of comprehensive international regulations specifically tailored to medical tourism complicates matters and makes it challenging to hold providers accountable for unethical practices. Some key regulatory challenges include:

1. Quality and safety concerns: Different countries have varying standards of healthcare quality and safety regulations. Patients who travel for medical procedures may face risks associated with inadequate healthcare facilities, unqualified healthcare providers, or substandard medical equipment. Creating international standards for patient safety and quality of care is essential.

2. Legal liabilities: Medical tourists may find themselves in a legal limbo when seeking recourse for medical complications or malpractice. The complex jurisdictional issues involved in medical tourism make it difficult for patients to hold negligent healthcare providers accountable.

3. Ethical considerations: Ethical dilemmas can arise when medical tourists receive procedures that are considered illegal or unethical in their home country. For example, some countries may offer surrogacy services or experimental treatments that are restricted or prohibited elsewhere. Determining the ethical boundaries and ensuring patient autonomy and informed consent can be challenging.

4. Transparency and information asymmetry: Medical tourists often face challenges in accessing complete and accurate information about healthcare providers, treatments, and costs. There is a considerable information asymmetry between patients and providers, making it difficult for patients to make

well-informed decisions. Ensuring transparency and full disclosure of risks, benefits, and costs is crucial.

Ethical considerations in medical tourism

Medical tourism raises several ethical concerns that need to be addressed to protect the interests of patients and uphold ethical principles. Here are some key ethical considerations in medical tourism:

1. Equity and justice: Medical tourism can exacerbate healthcare disparities both within and between countries. Wealthier individuals may have greater access to medical tourism, creating a situation where those with financial resources can receive superior care while those without means are left behind. Ethical frameworks should aim to promote equitable access to healthcare and prevent exploitation.

2. Informed consent: Informed consent is a fundamental ethical principle that requires patients to have a comprehensive understanding of the risks, benefits, alternatives, and limitations of a medical procedure. In the context of medical tourism, ensuring that patients have access to unbiased and complete information is crucial for obtaining truly informed consent.

3. Patient exploitation: The potential for patient exploitation is significant in medical tourism. Patients may face pressure to undergo unnecessary procedures or may be subjected to treatment options that are not evidence-based. It is crucial to protect vulnerable patients from being exploited and ensure that their best interests are prioritized over financial gains.

4. Cultural sensitivity: Medical tourists often encounter different cultural norms and practices in the host country. Healthcare providers must be culturally sensitive and respectful, ensuring that patients' cultural beliefs and values are taken into account in decision-making and treatment plans.

Regulatory and ethical solutions

Addressing the regulatory and ethical challenges in medical tourism requires a multifaceted approach involving international collaboration, policy development, and ethical guidelines. Some potential solutions include:

1. International collaboration: Collaborative efforts between countries can help establish a set of universally accepted standards and guidelines for medical tourism. This may involve the creation of regulatory bodies or the adaptation of existing frameworks such as the World Health Organization's (WHO) Global Code of Practice on the International Recruitment of Health Personnel.

2. Accreditation and certification: The establishment of accreditation and certification programs specific to medical tourism can help ensure that healthcare providers meet certain quality standards. This can provide patients with the reassurance that they are receiving care from reputable and competent providers.

3. Enhanced transparency and information dissemination: Improving transparency in medical tourism can be achieved through the development of centralized databases or registries that provide comprehensive and standardized information about healthcare providers, treatments, outcomes, and costs. This would enable patients to make well-informed decisions and compare options across different countries.

4. Ethical guidelines and training: Developing and disseminating ethical guidelines that address the unique challenges of medical tourism can help guide healthcare providers and organizations in upholding ethical practices. Training programs can also be implemented to enhance cultural sensitivity and ensure that healthcare professionals are equipped with the necessary skills to navigate the ethical complexities of medical tourism.

5. Patient advocacy and support: Establishing patient advocacy organizations or support networks can empower medical tourists and provide them with the necessary resources to navigate ethical dilemmas, legal issues, and post-treatment care. These organizations can also play a crucial role in raising awareness about the risks and benefits of medical tourism.

It is important to note that while regulatory and ethical measures can help mitigate some of the challenges associated with medical tourism, they may not completely eliminate risks or address underlying systemic issues. Continued collaboration, evaluation, and adaptation of these measures are essential to ensure the ethical and responsible growth of the medical tourism industry.

Summary

Regulating and ensuring ethical practices in medical tourism is a complex yet crucial task. The regulatory challenges include addressing quality and safety concerns, legal liabilities, ethical considerations, and information asymmetry. Ethical considerations in medical tourism involve equity and justice, informed consent, patient exploitation, and cultural sensitivity. To tackle these issues, international collaboration, accreditation, enhanced transparency, ethical guidelines, and patient advocacy are essential. By implementing these measures, the aim is to promote ethical and responsible medical tourism that prioritizes patient welfare and fosters a more equitable and just healthcare landscape.

Humanitarian Intervention and Health

Ethical Dilemmas in Humanitarian Aid Provision

Introduction

Humanitarian aid provision plays a crucial role in addressing the needs of vulnerable populations affected by natural disasters, conflict, and other emergencies. However, this noble endeavor is often accompanied by complex ethical dilemmas that require careful consideration. In this section, we will explore some of the key ethical dilemmas inherent in humanitarian aid provision, along with potential solutions and thought-provoking examples.

The Principle of Utility

The principle of utility, often associated with consequentialist ethical theories, holds that the right course of action is the one that maximizes overall well-being or happiness. In the context of humanitarian aid provision, this principle guides decision-making regarding resource allocation and service delivery. However, tensions can arise when trying to balance the needs of different individuals or groups.

Ethical Dilemma 1: Resource Allocation

One common ethical dilemma in humanitarian aid provision revolves around resource allocation. When resources are limited and the demand for aid is high, aid organizations must prioritize certain populations or regions over others. This raises questions of fairness and justice. How can aid organizations allocate resources in a way that is equitable and maximizes the overall impact?

Example: Triaging Medical Care   During a humanitarian crisis, medical services are often overwhelmed, and difficult decisions must be made regarding who receives care. This is particularly evident in triage situations, where medical professionals must decide which patients will receive treatment based on the severity of their condition and the likelihood of survival. Ethical guidelines, such as the concept of "doing the greatest good for the greatest number," can help inform these decisions. However, it remains a challenging task to balance the need for efficiency and fairness in resource allocation.

Ethical Dilemma 2: Neutrality

The principle of neutrality is a cornerstone of humanitarian aid provision, emphasizing that aid must be provided without discrimination based on factors such as nationality, ethnicity, or political affiliation. However, in certain situations, maintaining neutrality can become problematic when conflicts or political dynamics intersect with the delivery of aid.

Example: Negotiating Access In conflict zones, aid organizations often face challenges in negotiating access to affected populations. Non-state armed groups may demand control over aid distribution or impose conditions that compromise the principle of neutrality. Humanitarian actors must carefully navigate these dilemmas, aiming to balance the imperative to provide assistance with the need to maintain ethical standards. Balancing the safety of aid workers, maintaining relationships with local actors, and ensuring the provision of aid to those who need it most requires careful ethical considerations.

Ethical Dilemma 3: Cultural Sensitivity

Cultural sensitivity is crucial in humanitarian aid provision, as it respects the values, beliefs, and customs of the affected populations. However, cultural norms may sometimes clash with humanitarian principles and lead to ethical dilemmas.

Example: Female Empowerment In some contexts, traditional gender roles and discriminatory practices may inhibit the empowerment of women and girls. Humanitarian organizations face the challenge of balancing the respect for local cultures with their responsibility to promote gender equality and ensure the well-being of all individuals. Navigating this ethical dilemma requires a nuanced understanding of cultural contexts and collaboration with local stakeholders.

Ethical Dilemma 4: Short-term vs. Long-term Impact

Humanitarian aid provision often focuses on meeting immediate needs during emergencies. However, prioritizing short-term relief can sometimes undermine long-term development goals. Striking the right balance between immediate assistance and sustainable rehabilitation is an ethical challenge for humanitarian practitioners.

Example: Sustainable Food Aid Programs   In the aftermath of a natural disaster or conflict, providing food aid is essential. However, relying solely on external food aid can create dependency and hinder the development of local agricultural systems. Achieving a balance between immediate hunger relief and supporting long-term food security requires a holistic approach that engages local communities, promotes self-sufficiency, and recognizes the importance of long-term development.

Conclusion

Ethical dilemmas in humanitarian aid provision are unavoidable given the complexities of the contexts in which aid is delivered. Balancing the principles of utility, neutrality, cultural sensitivity, and long-term impact presents significant challenges. However, by engaging in critical reflection, consulting ethical guidelines, and working collaboratively with local actors, humanitarian organizations can navigate these dilemmas and strive to provide effective, ethical, and sustainable support to those in need. It is through this ongoing ethical dialogue that the humanitarian community can continuously improve its practices and uphold the rights and dignity of affected populations.

Medical Neutrality and Access to Healthcare in Conflict Zones

In conflict zones around the world, the provision of healthcare is often severely disrupted, leaving civilians without access to life-saving medical services. Medical neutrality is a fundamental principle that plays a crucial role in ensuring that healthcare is provided impartially and without discrimination in these challenging circumstances. This section will explore the concept of medical neutrality and discuss the ethical considerations surrounding access to healthcare in conflict zones.

Understanding Medical Neutrality

Medical neutrality is a principle rooted in the laws of armed conflict and human rights. It upholds the idea that medical professionals and facilities must be protected and allowed to provide healthcare services to all, regardless of their political affiliation, nationality, or ethnicity. This principle is enshrined in international humanitarian law, particularly the Geneva Conventions, and aims to safeguard the rights and wellbeing of individuals affected by armed conflicts.

To maintain medical neutrality, three key elements need to be upheld:

1. The prohibition of attacks on healthcare facilities, personnel, and transports.

2. The respect for the wounded, sick, and medical personnel.

3. The impartial provision of healthcare to all, without discrimination.

Challenges to Medical Neutrality

Despite the clear guidelines set forth by international law, medical neutrality often faces significant challenges in conflict zones. These challenges include:

+ Deliberate attacks on healthcare facilities and personnel: Parties involved in conflicts may intentionally target medical facilities and workers as a strategy to weaken the opposition or instill fear among civilians. Such attacks not only violate medical neutrality but also result in devastating consequences for the affected communities.

+ Obstruction of humanitarian aid: Governments or armed groups may hinder the delivery of medical supplies and humanitarian aid to areas affected by conflict. This obstruction limits access to essential healthcare services, putting the lives of civilians at risk.

+ Forced recruitment of healthcare professionals: In some conflicts, medical professionals are coerced into providing care exclusively to combatants, denying access to healthcare for civilians. This violates the principle of medical neutrality and obstructs the provision of impartial care.

+ Security risks for healthcare personnel: Medical personnel working in conflict zones often face significant security risks, including threats of violence, abduction, and even killings. These risks discourage healthcare workers from providing essential services, further limiting access to healthcare.

Ethical Considerations

Access to healthcare is a basic human right, even in times of conflict. Thus, ensuring medical neutrality and facilitating access to healthcare in conflict zones are key ethical imperatives. Several ethical considerations arise in this context:

+ Non-discrimination: Upholding the principle of medical neutrality requires that healthcare be provided without discrimination to all individuals, regardless of their affiliation or status. This includes combatants and civilians, refugees, and internally displaced persons.

+ Prioritization of need: In conflict zones, resources are often limited, and difficult decisions must be made regarding the allocation of healthcare services. Ethical frameworks emphasize the importance of prioritizing those in greatest need, including the wounded, children, pregnant women, and those with chronic illnesses.

+ Informed consent: In conflict settings, obtaining informed consent for medical interventions can be challenging due to disrupted communication channels and the traumatized conditions of individuals. Ethical guidelines stress the importance of striving for informed consent to the extent possible, while acknowledging the unique challenges present in conflict zones.

+ Protection of healthcare workers: Ensuring the safety and well-being of healthcare workers is paramount. Ethical considerations include providing adequate security measures, training, and support to enable healthcare professionals to fulfil their duty to provide care in conflict zones.

+ Collaboration and coordination: Addressing the complex healthcare needs in conflict zones requires collaboration and coordination among various stakeholders, including humanitarian organizations, governments, and local communities. Ethical frameworks emphasize the importance of promoting cooperation to ensure access to healthcare services.

Case Study: Syria

The conflict in Syria serves as a stark illustration of the challenges faced in maintaining medical neutrality and ensuring access to healthcare in a protracted conflict. Throughout the conflict, healthcare facilities have been deliberately targeted, leading to the destruction of hospitals and the killing and injuring of healthcare workers. Moreover, access to healthcare services has been significantly limited due to the obstruction of humanitarian aid and the forced displacement of populations.

In response to these challenges, various organizations, such as Médecins Sans Frontières (Doctors Without Borders) and the World Health Organization, have worked tirelessly to provide healthcare services and advocate for the protection of medical neutrality. These efforts highlight the importance of collaboration between humanitarian organizations, governments, and local actors to address the ethical dilemmas and promote access to healthcare in conflict zones.

Conclusion

The principle of medical neutrality is a cornerstone of providing healthcare in conflict zones. Upholding this principle ensures that all individuals, regardless of their affiliation or status, have access to essential medical services. However, numerous challenges, including attacks on healthcare facilities, obstruction of humanitarian aid, and security risks for healthcare workers, continue to impede the implementation of medical neutrality in conflict zones. Efforts to protect medical neutrality and promote access to healthcare require collaboration among various stakeholders, adherence to ethical considerations, and a commitment to upholding the human right to health even in the most challenging circumstances.

Human Rights and Health in Refugee Populations

The global refugee crisis has become a pressing humanitarian issue, with millions of individuals and families displaced from their homes due to conflict, persecution, or other dangerous circumstances. The plight of refugees raises significant ethical questions regarding their rights and access to healthcare. This section will explore the intersection of human rights and health in refugee populations, discussing the challenges they face and the ethical considerations involved.

Understanding Refugee Populations

Before delving into the ethical issues, it is essential to gain a deeper understanding of refugee populations. According to the United Nations High Commissioner for Refugees (UNHCR), a refugee is someone who has been forced to flee their country due to a well-founded fear of persecution based on race, religion, nationality, political opinion, or membership in a particular social group. Refugees often face extreme hardships, including the loss of livelihoods, separation from families, and exposure to violence and trauma.

Refugees are particularly vulnerable and require special attention and protection, including access to essential services such as healthcare. However, the reality is that many refugees live in overcrowded camps or informal settlements with limited access to adequate healthcare facilities. These challenging living conditions, coupled with the trauma and stress of displacement, place refugees at a higher risk of physical and mental health problems.

Human Rights and Health

The principles of human rights provide a foundation for understanding the ethical dimensions of providing healthcare to refugees. Human rights, as enshrined in international treaties and conventions, emphasize the inherent dignity and worth of every individual, regardless of their nationality or immigration status. Healthcare is considered a fundamental human right, as recognized by the Universal Declaration of Human Rights and other international agreements.

In the context of refugee populations, the right to health encompasses access to quality healthcare services, including preventive, curative, and rehabilitative care. It also includes the right to essential medicines, sanitation, safe drinking water, and other conditions necessary for good health. Protecting and promoting the right to health for refugees is crucial not only for ethical reasons but also for public health concerns, as the health of refugees directly impacts the health of host communities.

Challenges in Ensuring Health Rights for Refugees

While the principles of human rights establish a framework for the provision of healthcare to refugees, numerous challenges complicate the actual realization of these rights. The following challenges are worth highlighting:

1. **Limited access to healthcare services:** Refugees often face barriers in accessing healthcare services due to factors such as language barriers, lack of documentation, and unfamiliarity with the healthcare system in the host country. Moreover, healthcare facilities in host countries may be overwhelmed or ill-equipped to handle the influx of refugees, leading to long waiting times and inadequate care.

2. **Mental health needs:** Displacement and the traumatic experiences that accompany forced migration can have severe mental health consequences for refugees. However, mental health services are often scarce, under-resourced, and stigmatized. Many refugees may not receive the necessary support for coping with trauma, anxiety, depression, and other mental health conditions.

3. **Vulnerable groups within refugee populations:** Certain subgroups within refugee populations, such as women, children, the elderly, and individuals with disabilities, face heightened vulnerabilities and specific healthcare needs. Ensuring their access to appropriate and culturally sensitive healthcare services is essential but challenging.

4. **Healthcare disparities and discrimination:** Refugees may experience discrimination and marginalization within healthcare settings, affecting their ability to receive fair and equitable treatment. Healthcare providers must be mindful of cultural and language barriers, address implicit biases, and provide care that respects the dignity and rights of refugees.

Addressing these challenges requires a multi-faceted approach that involves collaboration between governments, international organizations, healthcare professionals, and civil society. An ethical framework for ensuring health rights for refugees should encompass the following principles:

1. **Equity and non-discrimination:** Healthcare services provided to refugees should be equitable, ensuring that all individuals have equal access to care based on their healthcare needs rather than their immigration status or nationality. This principle requires addressing language barriers, providing translation services, and culturally sensitive care.

2. **Participation and empowerment:** Meaningful participation of refugees in decisions affecting their health and well-being is essential. Involving refugees in the design and implementation of healthcare programs can help ensure that their unique needs and perspectives are considered. Empowering refugees to take an active role in their healthcare promotes autonomy and dignity.

3. **Collaboration and solidarity:** The responsibility to protect and promote the health rights of refugees rests not only with host countries but also with the international community. Collaboration and solidarity between governments, NGOs, and humanitarian organizations are critical for addressing the complex health challenges faced by refugees.

4. **Trauma-informed care:** Recognizing and addressing the trauma experienced by refugees is crucial in providing effective healthcare. Healthcare professionals should receive training in trauma-informed care to ensure that they can respond appropriately to the unique psychological and emotional needs of refugees.

Real-World Example: Providing Healthcare to Syrian Refugees

The Syrian refugee crisis serves as a real-world example to illustrate the ethical considerations and challenges in providing healthcare to refugee populations. Since the outbreak of the Syrian civil war in 2011, millions of Syrians have been displaced, seeking refuge in neighboring countries and beyond.

Countries such as Jordan, Lebanon, Turkey, and Iraq, which have hosted a significant number of Syrian refugees, have faced tremendous strain on their healthcare systems. Overcrowded hospitals, limited resources, and language barriers often result in substandard healthcare services for refugees. Moreover, the mental health needs of Syrian refugees, especially those who have experienced trauma and loss, are frequently overlooked due to the lack of specialized services.

International organizations and NGOs have played a crucial role in supporting healthcare services for Syrian refugees. These organizations provide funding, medical supplies, and technical assistance to strengthen the capacity of local healthcare systems. They also facilitate training programs for healthcare providers to enhance their skills in addressing the specific needs of refugees.

Conclusion

The health and well-being of refugee populations are intimately tied to their human rights. Despite the challenges, efforts to ensure the provision of healthcare services to refugees have the potential to not only alleviate suffering but also promote social justice and solidarity. By upholding the principles of equity, participation, collaboration, and trauma-informed care, we can work towards a future where every refugee has the opportunity to enjoy the right to health and wellness. The ethical imperative to support refugees in their quest for better health extends beyond borders, requiring a collective commitment to justice and compassion.

Ethical Issues in End-of-Life Care

Advance Directives and End-of-Life Decision-making

Living Wills and Healthcare Power of Attorney

In this section, we will explore the importance of living wills and healthcare power of attorney in making healthcare decisions for individuals who are unable to do so themselves. We will discuss the ethical considerations surrounding these legal documents and provide guidance on how to navigate the complexities of end-of-life care.

Background and Importance

As we age or face serious medical conditions, there may come a time when we are unable to make decisions about our own medical treatment. In such situations, it is crucial to have a plan in place that ensures our wishes are respected and that healthcare decisions align with our values and beliefs. This is where living wills and healthcare power of attorney come into play.

A living will, also known as an advance directive, is a legal document that allows individuals to express their preferences regarding medical treatment and end-of-life care. It provides instructions about the use of life-sustaining treatments such as resuscitation, mechanical ventilation, or artificial nutrition and hydration. By outlining our wishes in a living will, we can have a say in our healthcare decisions, even when we are incapacitated.

On the other hand, healthcare power of attorney, also known as a healthcare proxy or durable power of attorney for healthcare, designates a person (the healthcare agent or proxy) to make medical decisions on behalf of the individual who is unable to do so. This person is entrusted to make decisions that align with the individual's values and preferences, following the guidance provided in the living will, if one exists. It is crucial to select a trusted person as the healthcare agent, someone who understands and respects our values and will advocate for our best interests.

Both living wills and healthcare power of attorney empower individuals to retain control and autonomy over their medical treatment when they are no longer able to participate actively in decision-making. These legal documents provide peace of mind for individuals and their families, ensuring that their wishes are respected and reducing the burden of decision-making during challenging times.

Ethical Considerations

The use of living wills and healthcare power of attorney raises several ethical considerations. We will explore three key ethical principles that guide decision-making in this context: autonomy, beneficence, and justice.

Autonomy   Autonomy refers to an individual's right to make decisions about their own healthcare based on their values, beliefs, and preferences. With living wills, individuals can exercise their autonomy by explicitly stating their treatment preferences in advance. This allows them to maintain control over their medical care even when they are unable to communicate their wishes.

Healthcare power of attorney further upholds the principle of autonomy by allowing individuals to choose someone who understands and respects their values to make healthcare decisions on their behalf. By designating a healthcare agent, individuals exercise their autonomy in selecting someone who will advocate for their best interests when they are unable to do so.

Beneficence   Beneficence is the ethical principle that obligates healthcare professionals and individuals to act in ways that promote the well-being and best interests of the patient. Living wills and healthcare power of attorney support beneficence by providing clear guidance to healthcare providers, ensuring that medical treatments align with the individual's preferences and values.

When individuals create a living will, they provide healthcare professionals with valuable information that helps guide their decision-making. By adhering to the instructions outlined in the living will, healthcare providers can ensure that interventions are aligned with the individual's definition of a meaningful and dignified life.

Similarly, healthcare power of attorney allows the designated healthcare agent to make decisions in the best interests of the individual based on their values and preferences. This ensures that the patient's well-being remains the central focus, even when the individual is unable to voice their own wishes.

Justice   Justice pertains to fairness and the equitable distribution of resources and opportunities. In the context of living wills and healthcare power of attorney, justice requires that all individuals have access to these tools and that decisions are made without discrimination or bias.

It is essential to recognize that access to advance care planning and healthcare power of attorney may be limited due to various factors such as socioeconomic status, education, or cultural barriers. Efforts must be made to ensure that these resources are accessible to all individuals, regardless of their background or circumstances.

Healthcare providers and policymakers have a responsibility to promote justice by providing education, resources, and support to individuals and communities. By addressing barriers to accessing and understanding living wills and healthcare power of attorney, we can work towards a more equitable and just healthcare system.

Applying Living Wills and Healthcare Power of Attorney

Creating a living will and designating a healthcare power of attorney requires careful consideration and planning. Here are a few key steps to consider:

1.   **Discuss your wishes with loved ones**   Begin by having open and honest conversations with your loved ones about your values, beliefs, and healthcare preferences. It is crucial to express your desires clearly and ensure that your family members understand your intentions.

2.   **Consult with healthcare professionals**   Seek guidance from healthcare professionals, such as your primary care physician or an attorney specializing in healthcare law. They can provide valuable insights into the legal and medical aspects of advance care planning and help address any concerns or questions you may have.

3. **Create a living will**   Consult with an attorney to draft a living will that reflects your treatment preferences and end-of-life care wishes. Be specific about the medical interventions you would like to receive or refuse, considering different scenarios or medical conditions that may arise. Make sure to review and update your living will periodically to ensure it remains reflective of your current wishes.

4. Choose a trusted healthcare agent   Select a person you trust to serve as your healthcare agent or proxy. Discuss your values, beliefs, and treatment preferences with them, ensuring they understand your wishes and are willing to advocate for you in healthcare decision-making. Consider appointing an alternate agent in case the primary agent is unavailable or unable to fulfill their role.

5. Communicate and share your documents   Once you have created your living will and designated a healthcare agent, it is crucial to share these documents with your family members, primary care providers, and other relevant individuals. Keep copies of your documents in a safe and easily accessible location, and provide instructions on how to access them when needed.

Case Study: John's Journey

To illustrate the importance of living wills and healthcare power of attorney, let's consider the case of John, a 70-year-old man who has recently been diagnosed with a terminal illness. John wants to ensure that his treatment decisions align with his values and preferences while also alleviating the burden on his family members.

John consults with his primary care physician and an attorney specializing in healthcare law to better understand his options. He discusses his wishes with his children and selects his eldest son, Michael, as his healthcare agent.

With guidance from the attorney, John creates a living will that outlines his treatment preferences and end-of-life care wishes. He expresses his desire to avoid invasive treatments that offer little chance of improving his quality of life and instead focuses on palliative care and pain management.

John shares copies of his living will with Michael, his primary care physician, and his other children. He emphasizes the importance of open communication and encourages his family to discuss their feelings and concerns regarding his end-of-life care.

As John's health condition progresses, Michael closely collaborates with his father's healthcare team, ensuring that decisions are made in accordance with John's living will. He regularly communicates with his siblings, providing updates and seeking their input on treatment decisions when appropriate.

John passes away peacefully, knowing that his wishes were respected, and his family feels a sense of relief, having made decisions that aligned with their father's values.

Resources and Caveats

Creating a living will and designating a healthcare power of attorney can be complex processes. Here are some resources and caveats to consider:

Resources - Speak with an attorney specializing in healthcare law to ensure your legal documents meet all necessary requirements. - Consult organizations such as The Conversation Project or Aging with Dignity for guidance on advance care planning and end-of-life conversations. - Healthcare systems may have their own resources and protocols for advance care planning. Check with your primary care provider or local hospital for guidance.

Caveats - Laws regarding living wills and healthcare power of attorney may vary by jurisdiction. Consult with an attorney familiar with the laws in your area. - It is essential to keep your living will and healthcare power of attorney documents up to date. Review them periodically and make revisions as necessary to ensure they reflect your current wishes. - Communicating your wishes to your family is crucial. However, changes in circumstances or medical advances may require flexibility in decision-making, even with a living will in place.

Summary

Living wills and healthcare power of attorney provide individuals with the means to retain autonomy and control over their medical treatment, even when they are unable to make decisions themselves. These legal documents empower individuals to express their treatment preferences and designate a trusted healthcare agent, ensuring their values and beliefs are respected in healthcare decision-making.

By highlighting the ethical principles of autonomy, beneficence, and justice, we have explored the importance and application of living wills and healthcare power of attorney. While creating these documents requires careful consideration and planning, they offer peace of mind to individuals and their families, facilitating respectful and patient-centered end-of-life care.

Through open communication, education, and access to resources, we can work towards a society that values and supports advance care planning, creating a more ethical and compassionate approach to end-of-life care.

The Ethics of Withholding and Withdrawing Treatment

In the field of healthcare ethics, the decision to withhold or withdraw treatment is often a topic of great ethical importance and debate. This section explores the ethical considerations surrounding these decisions, including the principles and values that guide them, the challenges they present, and the ethical methods used to navigate them.

Understanding Withholding and Withdrawing Treatment

Withholding and withdrawing treatment refer to different scenarios in which healthcare providers make decisions not to initiate or to discontinue specific medical interventions for a patient. These decisions are typically made when the benefits of the treatment are deemed to be outweighed by the burdens and potential harm it may cause. Both ethical and legal considerations play a role in these decisions.

Principles and Values Guiding Withholding and Withdrawing Treatment

Several ethical principles and values guide the decision-making process when it comes to withholding or withdrawing treatment:

1. **Autonomy:** The principle of autonomy emphasizes the importance of respecting the patient's decision-making capacity and right to make decisions about their own healthcare. In the context of withholding or withdrawing treatment, autonomy requires that healthcare providers respect a patient's refusal of treatment or request to discontinue treatment.

2. **Beneficence and non-maleficence:** The principles of beneficence and non-maleficence require healthcare providers to act in the best interest of the patient and to do no harm. In the context of withholding or withdrawing treatment, these principles guide the assessment of whether the potential benefits of continued treatment outweigh the potential harm.

3. **Justice:** The principle of justice guides decisions about the fair distribution of healthcare resources. Healthcare providers must consider the allocation of resources in the context of withholding or withdrawing treatment, ensuring that resources are used in a just and equitable manner.

Challenges and Dilemmas

The decision to withhold or withdraw treatment can present several challenges and ethical dilemmas:

1. **Uncertainty:** Healthcare providers often face uncertainty when predicting the outcomes of withholding or withdrawing treatment. This uncertainty can make decision-making difficult and increase the ethical complexities involved.

2. **Value conflicts:** Different individuals involved in the decision-making process may have different values, leading to conflicts regarding what is the most appropriate course of action. For example, a patient's family may advocate for the continuation of treatment, while the healthcare team deems it non-beneficial or even harmful.

3. **Emotional impact:** Withholding or withdrawing treatment can have emotional implications for both patients and healthcare providers. Patients may experience fear, sadness, or a loss of hope, while healthcare providers may grapple with the weight of decision-making and the potential for moral distress.

Ethical Methods for Decision-making

Given the ethical complexities surrounding the withholding and withdrawing of treatment, various ethical methods have been developed to guide the decision-making process:

1. **Shared decision-making:** Shared decision-making involves collaboration between healthcare providers, patients, and their families, where all parties contribute to the decision-making process. This approach respects the autonomy of the patient while considering the expertise and values of the healthcare team.

2. **Ethics committees/consultation:** In challenging cases, healthcare institutions may establish ethics committees or seek ethical consultations. These committees and consultations provide a forum for interdisciplinary discussions, allowing for a more comprehensive analysis of the ethical dimensions involved.

3. **Advance care planning:** Advance care planning involves discussions between patients, their families, and healthcare providers to determine

preferences for future healthcare decisions. Engaging in advance care planning can help ensure that the patient's values and preferences are respected and followed when it comes to withholding or withdrawing treatment.

Case Example

Consider the case of an elderly patient with advanced dementia who has developed a serious infection. The healthcare team believes that aggressive treatment, such as intravenous antibiotics and hospitalization, is unlikely to provide significant benefits and may instead cause discomfort and distress to the patient. The patient's healthcare proxy, their daughter, insists on pursuing all possible treatments, as she believes there is still hope for improvement.

In this case, the principles of autonomy, beneficence, and non-maleficence come into play. The healthcare team must weigh the patient's autonomy and the daughter's perspective against the potential harm and burden of continued treatment. Through shared decision-making and considering the patient's previously expressed wishes, a collaborative decision can be reached that respects the patient's autonomy while aligning with the principles of beneficence and non-maleficence.

Additional Resources and Further Reading

1. Beauchamp, T. L., & Childress, J. F. (2019). *Principles of Biomedical Ethics.* Oxford University Press.

2. Sulmasy, D. P. (2017). *Methods in Medical Ethics.* Georgetown University Press.

3. American Society of Clinical Oncology. (2013). *Ethics in Oncology: Withholding and Withdrawing Therapy.* Retrieved from `https://www.asco.org/about-asco/people/leadership-structure/ethics/oncology-practice-committee/ethics-oncology`.

4. Wilkinson, D., Savulescu, J., & Skene, L. (Eds.). (2007). *A Companion to Bioethics.* John Wiley & Sons.

Physician-Assisted Death and Euthanasia

Physician-assisted death, also known as assisted suicide or euthanasia, is a highly controversial and emotionally charged topic in the field of health and wellness. It raises complex ethical, legal, and moral questions surrounding the role of healthcare professionals in aiding patients who wish to end their lives. In this section, we will explore the various perspectives on physician-assisted death, the ethical principles involved, the legal considerations, and the implications for healthcare providers.

Defining Physician-Assisted Death and Euthanasia

Physician-assisted death refers to the act of a medical professional intentionally providing the means or assistance to a patient in ending their own life. This may involve prescribing lethal doses of medication, providing information on self-administering a lethal drug, or directly administering a lethal substance. Euthanasia, on the other hand, is the act of intentionally ending a patient's life by a healthcare professional, usually by administering a lethal dose of medication.

It is important to distinguish between voluntary and involuntary euthanasia. Voluntary euthanasia involves patients who have given their explicit and informed consent to end their lives, typically due to unbearable suffering from a terminal illness. Involuntary euthanasia, on the other hand, is the act of ending a patient's life against their will, often without their knowledge or consent.

Ethical Considerations

The ethical debate surrounding physician-assisted death and euthanasia revolves around conflicting principles such as autonomy, beneficence, non-maleficence, and justice.

Autonomy: Autonomy refers to an individual's ability to make decisions regarding their own life and body. Advocates of physician-assisted death argue that respecting a patient's autonomy includes the right to control the timing and manner of their own death, especially in cases where they are suffering from incurable and unbearable pain. They argue that individuals have a fundamental right to die with dignity.

Beneficence and Non-Maleficence: The principles of beneficence (doing good) and non-maleficence (avoiding harm) are in conflict when it comes to physician-assisted death. Opponents argue that intentionally causing a patient's death contradicts the physician's duty to preserve life and do no harm. They argue that healthcare professionals should focus on providing palliative care and ensuring a comfortable, pain-free end-of-life experience for patients.

Justice: The principle of justice raises questions about access to physician-assisted death. Critics argue that legalizing physician-assisted death could disproportionately impact vulnerable populations, such as the elderly, disabled, or mentally ill, who might feel pressured to choose death due to societal, economic, or emotional factors. They emphasize the importance of ensuring equal access to quality end-of-life care for all patients.

Legal and Policy Considerations

Laws regarding physician-assisted death and euthanasia vary greatly across different jurisdictions. Some countries, such as the Netherlands, Belgium, Canada, Colombia, and parts of the United States, have legalized or decriminalized physician-assisted death under specific circumstances. Other countries, however, maintain strict laws prohibiting any form of euthanasia or assisted suicide.

Legal frameworks often require strict eligibility criteria to protect against abuse and ensure careful consideration of patient requests for physician-assisted death. Common requirements include a terminal illness prognosis, unbearable suffering, repeated requests, and evaluations by multiple healthcare professionals.

In countries where physician-assisted death is legal, policies are in place to safeguard the rights of patients and healthcare providers. These policies often include comprehensive informed consent procedures, mandatory waiting periods, thorough assessments of patient capacity, and reporting and oversight mechanisms to ensure compliance with the law.

Implications for Healthcare Providers

Physician-assisted death and euthanasia present unique challenges for healthcare providers. Balancing their duty to alleviate suffering and respect autonomy with their obligation to adhere to professional ethical guidelines can be ethically and emotionally demanding.

Healthcare providers must carefully consider their own personal beliefs, values, and ethical frameworks when confronted with requests for physician-assisted death. Some healthcare professionals may hold strong moral or religious objections to such practices, while others may feel it is their duty to alleviate suffering and honor patient autonomy.

Open communication and an interdisciplinary approach are essential in navigating discussions about physician-assisted death. It is crucial for healthcare providers to engage in thoughtful and respectful conversations with their patients, ensuring they are fully informed about all available options for end-of-life care, including palliative care and hospice services.

Furthermore, healthcare providers should be knowledgeable about the legal and ethical frameworks surrounding physician-assisted death in their jurisdiction. They should have access to resources and training to support them in making informed decisions and providing appropriate care to patients considering this option.

End-of-Life Care Alternatives

Physician-assisted death is often considered when patients perceive that they have no other options for alleviating their suffering. As healthcare providers, it is important to explore and promote alternatives to physician-assisted death.

Palliative care, for example, focuses on providing comfort and support to patients with life-threatening illnesses, aiming to enhance their quality of life by addressing physical, emotional, and spiritual needs. Integrating palliative care into the healthcare system and increasing access to these services can help alleviate suffering and provide patients with a sense of control and dignity.

Advance care planning and involvement of ethics committees can also contribute to informed decision-making and ensure that patients' values and wishes are respected. By engaging in discussions about end-of-life preferences early on, patients can express their desires regarding treatment options, resuscitation, and the use of life-sustaining interventions.

By promoting effective pain management, psychological support, and compassionate care, healthcare providers can strive to ensure that patients at the end of life experience the highest possible quality of life, even in the absence of physician-assisted death.

Case Study: The Oregon Death with Dignity Act

The Oregon Death with Dignity Act, passed in 1997, is one of the most well-known and studied laws permitting physician-assisted death. The act allows terminally ill Oregon residents, who meet specific criteria, to request lethal medications from their physicians. Since its implementation, thousands of patients have been prescribed lethal medications under this law.

The Oregon Death with Dignity Act includes safeguards such as mandatory waiting periods, requirements for patient capacity, evaluations by multiple physicians, and reporting mechanisms. Studies examining the implementation of this law have found that it has not led to an increase in non-voluntary euthanasia and has improved end-of-life care and access to hospice services for patients.

However, critics argue that such laws may have unintended consequences, such as a potential erosion of trust in the patient-physician relationship or the devaluation of vulnerable populations. Ongoing research and ongoing evaluation of the impact of these laws are crucial to ensure their ethical and appropriate implementation.

Conclusion

The topic of physician-assisted death and euthanasia raises complex ethical, legal, and moral questions. It requires careful consideration of competing principles such as autonomy, beneficence, non-maleficence, and justice. Healthcare providers must navigate these challenges with sensitivity, respecting patient autonomy while providing compassionate care and ensuring access to alternatives such as palliative care and hospice services. By engaging in open dialogue, understanding the legal and ethical frameworks, and promoting patient-centered care, healthcare professionals can navigate this complex and emotionally charged issue while upholding their professional obligations.

Palliative Care and Quality of Life

Ethical Considerations in Pain Management

Pain management is a crucial aspect of healthcare, and ethical considerations are paramount in ensuring that patients receive the appropriate care and support they need. In this section, we will explore the ethical dimensions of pain management, including the principles that guide decision-making, the challenges in balancing adequate pain control with the risk of misuse or addiction, and the importance of patient autonomy in pain treatment.

Principles of Ethical Decision-Making in Pain Management

Ethical decision-making in pain management is guided by several fundamental principles that help healthcare providers navigate the complex issues surrounding pain treatment. These principles include:

1. **Beneficence:** The principle of beneficence emphasizes the duty to promote the well-being and alleviate suffering in patients. In pain management, this principle calls for healthcare providers to prioritize the relief of pain and ensure that suffering is minimized to the greatest extent possible.

2. **Non-Maleficence:** The principle of non-maleficence emphasizes the duty to do no harm. In pain management, this principle requires healthcare providers to carefully assess the potential risks associated with pain medications or interventions and to take appropriate measures to minimize harm.

3. **Autonomy:** The principle of autonomy recognizes that patients have the right to make decisions about their own healthcare. In pain management, this principle highlights the importance of involving patients in decisions about their pain treatment, respecting their preferences, and obtaining informed consent.

4. **Justice:** The principle of justice pertains to the fair distribution of resources and the equitable access to pain management services. In pain management, this principle calls for healthcare providers to ensure that all patients, regardless of their socioeconomic status, gender, race, or other factors, have access to effective pain relief.

These principles provide a framework for ethical decision-making in pain management and guide healthcare providers in their efforts to provide compassionate and patient-centered care.

Balancing Pain Control and the Risk of Misuse or Addiction

One of the central ethical challenges in pain management is finding the right balance between adequate pain control and the potential risk of misuse or addiction to pain medications. This challenge is particularly relevant in the context of opioid medications, which are highly effective in relieving pain but also carry a significant risk of dependence and addiction.

Healthcare providers must exercise caution and use evidence-based guidelines when prescribing opioid medications. This includes conducting a thorough assessment of the patient's pain, closely monitoring the patient for signs of misuse or addiction, and implementing strategies to minimize the risk of harm.

Additionally, healthcare providers should prioritize the use of multimodal pain management approaches, which combine non-opioid medications, physical therapies, complementary and alternative therapies, and psychological interventions. This approach reduces reliance on opioids and may mitigate the risk of opioid-related harm.

Patient Autonomy in Pain Treatment

Respecting patient autonomy is essential in pain management. Patients should be actively involved in decisions about their pain treatment and have the right to make informed choices based on their preferences, values, and goals.

Healthcare providers should ensure that patients are adequately informed about the benefits, risks, and alternatives of various pain management options.

This includes providing information about the potential side effects, potential for addiction, and the availability of non-opioid alternatives.

Shared decision-making, in which healthcare providers collaborate with patients to develop a pain management plan that aligns with the patient's values and goals, is crucial in promoting patient autonomy. This approach fosters a therapeutic relationship built on trust and mutual respect.

Challenges and Solutions

Pain management presents several challenges in ethical decision-making. Some of the common challenges include:

1. **Under-treatment of Pain:** Patients may face barriers to accessing appropriate pain management, leading to under-treatment of pain. This can be due to various factors such as healthcare provider biases, inadequate pain assessment, or limited access to pain medications. Healthcare systems should strive to address these barriers and ensure that all individuals receive equitable and effective pain relief.

2. **Placebo Use:** There is an ongoing debate about the ethical use of placebos in pain management. While placebos may provide temporary relief for some patients, their use raises ethical concerns related to deception and the possibility of withholding effective treatment. Transparency in the use of placebos is crucial, and healthcare providers should consider alternative approaches that prioritize evidence-based and patient-centered care.

3. **End-of-Life Pain Management:** Pain management at the end of life requires particular attention to ethical considerations. In these situations, the principle of balancing pain control with the risk of harm becomes crucial. Healthcare providers must prioritize ensuring adequate pain relief while considering the potential for side effects and the patient's overall comfort.

4. **Equitable Access to Pain Management:** Disparities in access to pain management services can disproportionately affect vulnerable populations, including racial and ethnic minorities, individuals with limited financial resources, and those with limited education. Healthcare providers and policymakers must work together to address these disparities and ensure equitable access to pain relief.

These challenges can be addressed through comprehensive pain management education for healthcare providers, the development of evidence-based guidelines, and the implementation of policies that prioritize equitable access to pain relief.

Conclusion

Ethical considerations play a vital role in pain management, influencing decision-making, patient-provider relationships, and overall healthcare delivery. By adhering to the principles of beneficence, non-maleficence, autonomy, and justice, healthcare providers can navigate the complexities of pain management and ensure that patients receive compassionate, effective, and ethically sound care.

Through shared decision-making, a focus on multimodal approaches, and a commitment to addressing disparities in pain management, healthcare providers can foster a more ethical and patient-centered approach to pain treatment. This, in turn, can contribute to improved patient outcomes and a more equitable healthcare system.

Psychological and Spiritual Support in Palliative Care

In palliative care, the focus is not only on managing physical symptoms and providing medical treatment but also on addressing the psychological and spiritual needs of patients and their families. It recognizes that the end of life can be a challenging and transformative experience, and therefore, psychological and spiritual support plays a crucial role in promoting holistic well-being. This section explores the importance of psychological and spiritual support in palliative care, the ethical considerations involved, and strategies for providing effective support.

The Importance of Psychological Support

Psychological support aims to address the emotional, cognitive, and behavioral aspects of a patient's well-being. It helps individuals cope with the distress associated with their illness, enhance their quality of life, and improve their ability to communicate their needs and wishes.

1. Managing emotional distress: Palliative care patients often experience a range of emotions such as fear, anxiety, grief, and sadness. Psychological support provides a safe environment for patients to express their emotions and receive validation and understanding. It may involve individual counseling, support groups, or interventions such as cognitive-behavioral therapy to help patients develop coping strategies and enhance their emotional well-being.

2. Facilitating communication: Illness can sometimes lead to communication difficulties, making it challenging for patients to express their preferences or discuss their concerns. Psychological support can help patients and their families navigate these challenges by improving their communication skills, providing guidance on difficult conversations, and promoting open and honest dialogue.

3. Enhancing quality of life: Psychological support aims to enhance the overall quality of life for patients by fostering a sense of purpose, meaning, and hope. It may involve interventions such as life review, legacy work, or existential therapy that allow patients to reflect on their life experiences, find meaning in their illness, and promote a sense of dignity and closure.

The Role of Spiritual Support

Spiritual support recognizes that patients may have spiritual or existential concerns related to their illness and mortality. It focuses on providing comfort, meaning, and hope by addressing the spiritual needs of individuals, regardless of religious affiliation.

1. Addressing existential concerns: Facing mortality often raises profound existential questions about the purpose of life, the nature of suffering, and the search for meaning. Spiritual support offers a space for patients to explore these concerns, find answers within their own beliefs or values, and find solace and peace.

2. Providing comfort and hope: Spiritual support can offer a source of comfort and hope in times of distress. It may involve providing presence and companionship, conducting religious or spiritual rituals, or facilitating connections with clergy or spiritual leaders who can address the specific spiritual needs of the patient.

3. Accompanying the journey: Spiritual support acknowledges that the end-of-life journey is unique for each individual. It involves being present with patients, actively listening to their stories, and accompanying them in their spiritual and existential explorations.

Ethical Considerations

The provision of psychological and spiritual support in palliative care raises several ethical considerations:

1. Respect for autonomy: It is essential to respect the autonomy and personal beliefs of patients when providing psychological and spiritual support. Care must be taken not to impose any specific religious or spiritual beliefs on patients but rather to create an inclusive and supportive environment that respects their values and choices.

2. Confidentiality: Psychological and spiritual support often involves deeply personal and sensitive information. Healthcare providers must prioritize patient confidentiality and ensure that all information shared within this context is kept private and only shared with the patient's explicit consent.

3. Multidisciplinary collaboration: Effective psychological and spiritual support requires collaboration among different healthcare professionals, including psychologists, counselors, social workers, and chaplains. Ethical considerations include maintaining clear communication, respecting professional boundaries, and ensuring that the patient's best interests are at the forefront of care.

Strategies for Providing Effective Support

1. Comprehensive assessment: A comprehensive assessment of the patient's psychological and spiritual needs should be conducted regularly. This assessment may involve standardized measures, clinical interviews, and discussions with the patient and their family members.

2. Collaborative care planning: Following the assessment, an individualized care plan should be developed collaboratively with the patient, their family, and the interdisciplinary team. This plan should address the specific psychological and spiritual needs identified and outline the interventions and support services to be provided.

3. Training and education: Healthcare professionals involved in palliative care should receive specialized training and education in psychological and spiritual support. This training should focus not only on theoretical knowledge but also on developing practical skills for effective communication, active listening, and providing support in a compassionate and empathic manner.

4. Integration of services: Psychological and spiritual support should be integrated into the overall care provided to patients. This requires close collaboration between different healthcare professionals, effective communication, and regular updates on the patient's progress and needs.

5. Ongoing evaluation and feedback: Regular evaluation of the psychological and spiritual support services provided is crucial to ensure their effectiveness and address any potential gaps or challenges. Feedback from patients, families, and healthcare professionals should be sought and incorporated into continuous quality improvement efforts.

As palliative care continues to evolve, the recognition of the importance of psychological and spiritual support is crucial. By addressing the holistic needs of patients, healthcare professionals can provide comprehensive and compassionate care that respects the dignity and autonomy of individuals at the end of life.

Cultural Perspectives on Death and Dying

Death and dying are universal experiences that are deeply rooted in cultural beliefs, values, and practices. In different cultures around the world, there are diverse perspectives on death and dying, rituals associated with death, and ways of grieving and honoring the deceased. Understanding cultural perspectives on death and dying is essential for healthcare professionals to provide culturally sensitive and respectful end-of-life care.

Cultural Beliefs and Attitudes toward Death

Cultural beliefs about death often shape how individuals and communities understand and approach the end of life. For example:

+ **Eastern cultures:** Many Eastern cultures, such as those influenced by Buddhism, Hinduism, and Confucianism, believe in the concept of reincarnation or rebirth. Death is seen as a natural part of the cycle of life, and the focus is often on achieving a peaceful and harmonious transition to the afterlife.

+ **Western cultures:** In Western cultures, death is often viewed as the end of life, with an emphasis on individualism and the pursuit of personal autonomy. Many Western societies also have religious traditions that shape beliefs and practices surrounding death.

+ **Indigenous cultures:** Various Indigenous cultures have unique perspectives on death and dying. For example, some Native American tribes believe in the interconnectedness of all things and may view death as a continuation of life in a different form.

+ **African cultures:** African cultures have diverse beliefs about death, often related to ancestral worship and the continuation of familial bonds beyond death. Rituals and ceremonies play a significant role in honoring the deceased and facilitating their journey to the afterlife.

It is important to note that these are just a few examples, and cultural perspectives on death can vary widely even within a specific culture or region. Healthcare professionals should approach end-of-life care with cultural humility and respect for the diverse beliefs and attitudes of individuals and communities.

Death Rituals and Funeral Practices

Funeral rituals and practices vary greatly across cultures and are deeply rooted in tradition, religious beliefs, and cultural customs. These rituals serve several purposes, including honoring the deceased, comforting the bereaved, and guiding the transition of the deceased to the afterlife. Here are some examples of cultural practices related to death and funerals:

+ **Asian cultures:** In many Asian cultures, such as Chinese, Japanese, and Korean, funeral rites often involve elaborate ceremonies and rituals. Ancestor worship and the veneration of deceased family members play a significant role. Practices may include the burning of incense, offering food and prayers, and elaborate mourning periods.

+ **Mexican culture:** In Mexican culture, Day of the Dead (Dia de los Muertos) is a significant annual celebration honoring deceased loved ones. Families create colorful altars (ofrendas) adorned with photographs, favorite foods, and mementos of the deceased. The celebration involves music, dancing, and storytelling to remember and celebrate the lives of the departed.

+ **Islamic culture:** Islamic funeral practices typically involve simple burials conducted as soon as possible after death. There are specific rituals surrounding the preparation of the body, including washing and wrapping it in a white shroud. Prayers are recited, and the body is buried in a simple grave, ideally facing Mecca.

+ **Native American traditions:** Native American tribes have diverse funeral customs and rituals. Some tribes practice sky burials, where the deceased are placed on an elevated platform or exposed to the elements. Others may conduct ceremonies involving drumming, chanting, and the use of sacred herbs to guide the spirit of the deceased to the afterlife.

These examples highlight the rich diversity of death rituals and funeral practices worldwide. Healthcare professionals should be aware of and respect these cultural practices when providing end-of-life care to individuals from different cultural backgrounds.

Grief and Mourning Practices

Grief is a natural response to the loss of a loved one and is influenced by cultural and social contexts. Different cultures have unique ways of expressing and coping with grief. Here are some examples of cultural practices related to grief and mourning:

+ **Irish wake:** In Irish culture, a wake is a gathering held before a funeral where family and friends come together to mourn the deceased. It is often characterized by storytelling, music, and sharing fond memories of the departed.

+ **African-American traditions:** African-Americans have rich traditions of gathering and mourning collectively. Practices such as "homegoing" ceremonies, community vigils, and soul food dinners provide opportunities for the community to come together, support each other, and celebrate the life of the deceased.

+ **Tibetan Buddhist tradition:** In Tibetan Buddhist tradition, a mourning period of 49 days is observed after the death of a loved one. During this time, prayers and rituals are performed to guide the deceased through the intermediate state between death and rebirth.

+ **Jewish tradition:** In Jewish culture, mourning practices include a period of intense grief called shiva, lasting seven days. During shiva, family and friends gather in the home of the deceased to offer condolences, recite prayers, and support the bereaved.

Grief and mourning practices can also vary within cultures based on factors such as religious beliefs, regional customs, and individual preferences. Healthcare professionals should be sensitive to these variations and provide support and resources that align with the cultural and individual needs of the bereaved.

Challenges and Considerations in Cultural Perspectives on Death and Dying

While cultural perspectives on death and dying enrich our understanding of diverse beliefs and practices, there are also challenges and ethical considerations to navigate. Some of these challenges include:

+ **Communication barriers:** Language and cultural differences can create barriers in effectively communicating with patients and their families about end-of-life care preferences and decisions. Healthcare professionals should employ interpreters when necessary and use culturally appropriate communication strategies to ensure understanding and respect.

+ **Conflicting values and beliefs:** Cultural perspectives on death may conflict with medical norms and practices. For example, some cultural and religious beliefs may prohibit certain medical interventions or end-of-life decisions. Healthcare professionals must navigate these conflicts with sensitivity, respect, and adherence to ethical principles.

+ **Cultural stereotypes and biases:** Stereotypes and biases can influence how healthcare professionals perceive and respond to patients' cultural practices and beliefs surrounding death and dying. It is essential to challenge these biases and provide culturally sensitive care that honors individuals' cultural values and preferences.

+ **Legal and ethical considerations:** There may be legal and ethical considerations when accommodating cultural practices surrounding death and dying. For example, certain cultural practices related to burial or organ donation may conflict with legal regulations. Healthcare professionals must balance cultural sensitivity with legal and ethical obligations.

Culturally competent care requires healthcare professionals to actively seek knowledge, reflect on their biases, and adapt their practice to provide respectful and person-centered care that aligns with patients' cultural perspectives on death and dying.

Conclusion

Cultural perspectives on death and dying greatly influence how individuals and communities experience and navigate the end of life. Recognizing and honoring these cultural beliefs, rituals, and practices is vital for providing culturally sensitive and holistic end-of-life care.

In this section, we explored various cultural perspectives on death and dying, including beliefs, funeral practices, grief and mourning rituals, and the challenges involved in navigating cultural differences in end-of-life care. Healthcare professionals must strive for cultural humility, actively learn about diverse cultural practices, and integrate this knowledge into their practice to ensure equitable, respectful, and person-centered care for all individuals and communities.

By acknowledging and valuing cultural perspectives on death and dying, we can promote dignity, compassion, and support during life's final chapter, creating a more inclusive and ethically sound approach to end-of-life care in a diverse and multicultural world.

Care for Dying Children and Neonates

Ethical Issues in End-of-Life Care for Children

End-of-life care for children is a sensitive and complex area that raises unique ethical considerations. When a child is facing a life-limiting illness or condition, healthcare professionals and families must navigate difficult decisions to ensure that the child's well-being and dignity are preserved. In this section, we will explore the ethical issues surrounding end-of-life care for children, including the principles of beneficence, autonomy, and justice, as well as the challenges associated with decision-making, pain management, and the involvement of parents and caregivers.

The Principle of Beneficence

The principle of beneficence, which emphasizes the promotion of well-being and the prevention of harm, is particularly crucial in end-of-life care for children. In this context, healthcare professionals have an ethical obligation to ensure that the child receives the best possible care to alleviate suffering and improve quality of life. However, determining what constitutes "beneficial" care can be challenging, as it requires careful consideration of the child's wishes, values, and cultural beliefs, as well as the prognosis and available treatment options.

One ethical dilemma that arises in end-of-life care for children is the use of palliative sedation. Palliative sedation involves intentionally reducing a child's consciousness to relieve refractory symptoms that cannot be adequately controlled through other means. While this practice is aimed at minimizing suffering, it raises concerns about the child's autonomy and dignity. Healthcare professionals must carefully balance the goals of symptom management with the child's right to be conscious and participate in decision-making, if possible.

The Principle of Autonomy

The principle of autonomy recognizes the child's right to make decisions about their care to the extent that they are capable. However, determining a child's capacity to make decisions can be challenging, particularly in the context of end-of-life care. Younger children may not possess the cognitive ability to fully understand their condition and the implications of different treatment options. In these cases, healthcare professionals must rely on surrogate decision-makers, typically parents or legal guardians, to make decisions in the child's best interests.

The involvement of parents in decision-making can also create ethical dilemmas, especially when there are disagreements about the appropriate course of action. In such situations, healthcare professionals must engage in open and honest communication with parents, respecting their perspectives while also considering the best interests of the child. Mediation and ethics consultations may be necessary to facilitate consensus and ensure that the child's welfare remains paramount.

The Principle of Justice

The principle of justice emphasizes fairness and equity in the allocation of resources and decision-making processes. In the context of end-of-life care for children, justice requires that all children have equal access to appropriate care, regardless of their socioeconomic status, ethnicity, or other demographic factors. However, systemic barriers and inequalities can hinder the achievement of justice in practice.

For example, some families may face financial hardships in accessing specialized pediatric palliative care services or evidence-based treatments. Healthcare professionals and policymakers have a responsibility to address these disparities and advocate for equitable access to resources and support. Collaboration with social workers, financial counselors, and community organizations can help ensure that families receive the necessary assistance to navigate the complexities of end-of-life care.

Pain Management and Symptom Relief

One of the primary goals of end-of-life care for children is to manage pain and other distressing symptoms effectively. The ethical imperative to alleviate suffering requires healthcare professionals to employ evidence-based practices for pain management, tailored to the child's individual needs and preferences. However, achieving optimal pain relief can be challenging, especially when balancing the potential risks and benefits of different medications and treatments.

One ethical dilemma in pain management is the use of opioids. While opioids can effectively relieve severe pain, there are concerns about their potential for addiction and respiratory depression. Healthcare professionals must carefully monitor and titrate the dosage of opioids to ensure adequate pain control while minimizing the risk of adverse effects. Additionally, open and transparent communication with parents about the benefits, risks, and alternatives to opioid medications is essential to obtain their informed consent and maintain trust.

Supporting Families and Caregivers

End-of-life care for children places a tremendous emotional burden on families and caregivers. Ethical considerations extend beyond the child's well-being to encompass the support and guidance necessary to help families navigate the complexities of decision-making, grief, and bereavement. Healthcare professionals should provide comprehensive psychosocial support, ensure clear and compassionate communication, and facilitate the involvement of spiritual and cultural resources as needed.

Additionally, actively involving parents and caregivers in care planning and decision-making processes is vital to uphold their autonomy and sense of agency. Collaborative decision-making models, such as shared decision-making, can empower parents to make informed choices that align with their values and goals for their child. By integrating families as partners in the care team, healthcare professionals can create an ethical and compassionate environment that supports families throughout their journey.

Ethical Considerations in Practice

When providing end-of-life care for children, healthcare professionals must be aware of their own values, biases, and emotions that may influence their decision-making and interactions with families. Self-reflection, ongoing education, and interdisciplinary collaboration are essential to enhance ethical competence and mitigate the risk of harm to the child and family.

Furthermore, it is important to recognize the limitations of medical interventions and embrace the value of compassionate presence and supportive care. Ethical end-of-life care for children involves acknowledging the child's uniqueness, preserving their dignity, and honoring their wishes to the fullest extent possible, while providing unwavering support to families during this difficult time.

Conclusion

Ethical issues in end-of-life care for children are complex and require a multifaceted approach that balances the principles of beneficence, autonomy, and justice. Healthcare professionals must navigate decision-making, pain management, and support for families with sensitivity and compassion. By recognizing the unique needs of children and families, fostering open communication, and advocating for equitable access to resources, we can ensure that end-of-life care for children upholds the principles of dignity, compassion, and respect.

Infant and Neonatal Palliative Care

Infant and neonatal palliative care is a specialized field that focuses on providing compassionate and supportive care for infants and newborns with life-limiting conditions or terminal illnesses. This branch of healthcare recognizes that some infants may not survive, and aims to enhance their quality of life and support their families during this difficult time.

Background

Infant and neonatal palliative care has emerged as a distinct discipline due to advancements in medical technology and the recognition that infants and newborns can experience pain and suffering. It acknowledges the unique needs of these young patients and their families, providing a holistic approach that addresses physical, emotional, and spiritual aspects of care.

Principles

The principles guiding infant and neonatal palliative care are rooted in the broader principles of palliative care, adapted specifically for this population. These principles include:

1. Communication and shared decision-making: Open and honest communication is crucial when discussing prognosis, treatment options, and end-of-life decisions with parents and caregivers. Shared decision-making involves actively involving families in the decision-making process, considering their values, beliefs, and cultural backgrounds.

2. Pain and symptom management: Neonates and infants can experience pain and discomfort, which can be effectively managed through the use of specific pharmacological and non-pharmacological interventions. It is essential to assess and address pain and symptoms promptly to provide comfort and improve the quality of life for these young patients.

3. Emotional and psychological support: Infant and neonatal palliative care recognizes the emotional and psychological needs of families, providing counseling, psychological support, and bereavement services. It acknowledges the grief and anticipatory grief experienced by parents and offers resources for coping and healing.

4. Continuity of care: Coordination and collaboration among healthcare professionals, including neonatologists, pediatricians, nurses, social workers, and chaplains, is critical to ensure seamless and integrated care for the infant and their family. Continuity of care also involves smooth transitions between hospital, home, and hospice settings, if applicable.

5. Cultural sensitivity: Infant and neonatal palliative care respects diverse cultural practices, beliefs, and customs. It strives to provide care that is culturally sensitive and responsive to the individual needs and preferences of families.

Challenges and Ethical Considerations

Infant and neonatal palliative care brings forth several challenges and ethical considerations:

1. Ethical dilemmas in decision-making: Decisions regarding the initiation or withdrawal of life-sustaining treatments, including resuscitation, mechanical ventilation, and surgery, can be complex and emotionally challenging. Healthcare providers must navigate these ethical dilemmas while prioritizing the best interests of the infant and respecting the wishes of the family.

2. Uncertainty surrounding prognosis: The prognosis for infants and neonates with life-limiting conditions may be uncertain, making decision-making even more challenging. Healthcare providers must communicate this uncertainty effectively, providing families with all available information to make informed decisions.

3. Resource allocation: Pediatric healthcare resources, including specialized neonatal intensive care units, are finite. Allocating these resources ethically and equitably raises questions about the distribution of care and the balance between providing intensive therapies and supporting palliative care services.

4. Ethical considerations in end-of-life care: End-of-life care decisions, such as withholding or withdrawing life-sustaining treatments, palliative sedation, and the consideration of euthanasia or medical aid in dying, require careful ethical analysis and consideration. These decisions must adhere to legal and ethical frameworks while respecting the ethical principles of beneficence, non-maleficence, autonomy, and justice.

Support and Resources

Families facing the challenges of infant and neonatal palliative care require comprehensive support services and resources. Some of these include:

1. Palliative care teams: Many healthcare institutions have interdisciplinary palliative care teams that specialize in neonatal and infant palliative care. These teams consist of professionals such as physicians, nurses, social workers, chaplains, and bereavement coordinators who can provide holistic support and guidance.

2. Support organizations: Several support organizations specifically cater to families navigating infant and neonatal palliative care. These organizations offer emotional support, counseling, financial assistance, and educational resources.

3. Bereavement services: Specialized bereavement programs and services are available to support families before and after the loss of their infant. These services often include counseling, support groups, and remembrance activities.

Case Study

Consider the case of Sarah, a newborn diagnosed with a severe congenital heart defect. The medical team informs her parents, John and Emily, about the prognosis and the limited treatment options available. With the support of the neonatal palliative care team, John and Emily explore various treatment approaches, understanding the potential risks and benefits. Ultimately, they decide to focus on providing Sarah with comfort care and optimizing her quality of life. The neonatal palliative care team provides ongoing support to the family, addressing their emotional needs, assisting in care planning, and facilitating effective communication with other healthcare providers.

Conclusion

Infant and neonatal palliative care recognizes the unique needs of infants and their families facing life-limiting conditions. By adopting the principles of open communication, pain management, emotional support, continuity of care, and cultural sensitivity, healthcare providers can enhance the quality of life for these young patients and their families. Despite the challenges and ethical considerations involved, specialized support services and resources exist to assist families through this difficult journey.

Parental Decision-making for Dying Children

When a child is facing a life-threatening illness or injury, parents are often confronted with the difficult task of making decisions regarding their child's medical care, including end-of-life decisions. This section explores the ethical considerations surrounding parental decision-making for dying children, aiming to provide guidance and support for parents in these challenging situations.

The Principle of Best Interest

At the heart of parental decision-making for dying children is the principle of best interest. This principle asserts that decisions should be made based on what is in the best interest of the child, taking into account their unique circumstances, values, and desires. However, defining what constitutes the child's best interest can be complex and subjective.

Parents must consider both the potential benefits and burdens of medical interventions for their child. They may need to weigh the potential for relief of suffering against the potential for negative side effects or prolonged pain. They may also need to consider the quality of life that their child will have if certain interventions are pursued.

It is important for parents to have open and ongoing discussions with the healthcare team to fully understand the medical options available, the risks and benefits associated with each option, and how each option aligns with the child's best interest.

Shared Decision-making

While parents have the legal authority to make medical decisions for their children, healthcare professionals play a crucial role in guiding and supporting parents in making informed decisions. Shared decision-making is a collaborative approach that involves healthcare professionals providing parents with accurate information, discussing the available options, and respecting the values and preferences of the parents.

Healthcare professionals should provide parents with a clear understanding of the child's medical condition, prognosis, and the potential outcomes of different treatment options. They should explain the risks and benefits in an understandable manner, ensuring that parents have the information they need to make well-informed decisions.

It is essential for healthcare professionals to create a supportive and empathetic environment, so that parents feel comfortable expressing their concerns, fears, and hopes. This helps to establish a trusting relationship and promotes shared decision-making.

Ethical Issues and Challenges

Parental decision-making for dying children is not without ethical challenges. Some of the key issues include:

1. **Decision-making under uncertainty:** The prognosis for a dying child may be uncertain, making decisions about treatment and end-of-life care particularly challenging. Parents may experience significant emotional distress and struggle with the uncertainty of outcomes.

2. **Cultural and religious beliefs:** Parents' cultural and religious beliefs may play a significant role in their decision-making process. Healthcare professionals should respect and incorporate these beliefs, ensuring that parents' values and preferences are considered.

3. **Balancing hope and realism:** Parents may cling to hope for a miracle or a cure, while healthcare professionals need to provide realistic expectations. Striking a balance between hope and realism is a delicate task that requires sensitivity and understanding on the part of healthcare professionals.

4. **Sustaining the parent-child bond:** Parents face the challenge of making decisions that not only provide the best medical care for their child but also allow them to spend meaningful time together and maintain a sense of normalcy. These decisions require careful consideration of the child's physical, emotional, and social needs.

5. **Dealing with guilt and regret:** Parents may experience guilt and regret regardless of the decisions they make. Healthcare professionals should offer emotional support and reassurance to parents, helping them navigate these challenging emotions.

Case Study: The Thompson Family

To further illustrate the complexities of parental decision-making for dying children, let's consider the case of the Thompson family. Their 8-year-old daughter, Emily, has been battling a rare and aggressive form of cancer for the past year. Despite multiple treatment modalities, Emily's condition has deteriorated, and her healthcare team believes that further treatment would be futile.

Mr. and Mrs. Thompson are faced with the difficult decision of whether to continue aggressive treatment or shift the focus to palliative care. They struggle with the fear of losing their daughter and the guilt associated with potentially "giving up" on her.

Emily's oncologist, Dr. Rodriguez, schedules a family meeting to discuss the prognosis and the available options. During the meeting, Dr. Rodriguez provides the Thompsons with a detailed explanation of Emily's condition, the potential outcomes of continued aggressive treatment, and the benefits of transitioning to palliative care. She empathizes with the Thompsons, acknowledging the magnitude of the decision they have to make.

After extensive discussions and soul-searching, the Thompsons ultimately decide to transition Emily to palliative care. They express their desire to focus on ensuring Emily's comfort and quality of life in her remaining time. Dr. Rodriguez supports their decision and ensures that the palliative care team will provide comprehensive support to manage Emily's symptoms and support the emotional well-being of the Thompsons.

Resources and Support

Parents who are facing the difficult task of making decisions for their dying child should be aware of the resources and support available to them. These may include:

+ **Hospital palliative care teams:** These teams specialize in helping families navigate the complexities of end-of-life care, providing physical, emotional, and spiritual support. They can assist in decision-making, ensure adequate pain and symptom management, and facilitate open and honest communication.

+ **Hospice organizations:** Hospice organizations offer comprehensive end-of-life care and support, both in the hospital and at home. They provide expert guidance, counseling, and a range of services to ensure that the child and family have the best possible quality of life during this time.

+ **Support groups and counseling services:** These resources can provide a safe space for parents to share their experiences, express their emotions, and find peer support. Counseling services can offer professional guidance to help parents navigate the emotional challenges they encounter.

+ **Ethics consultations:** If parents or healthcare professionals are unsure about the best course of action or face ethical dilemmas, ethics consultations can be sought. These consultations bring together an interdisciplinary team of experts who can provide guidance on complex ethical issues.

Conclusion

Parental decision-making for dying children is an extraordinarily difficult and emotionally charged process. It requires collaboration between parents and healthcare professionals and should be guided by the principle of best interest. This section has explored the ethical considerations, challenges, and resources involved in making such decisions, aiming to support parents during this incredibly challenging time. By providing guidance, empathy, and comprehensive support, healthcare professionals can assist parents in making decisions that prioritize the well-being and best interest of the dying child.

Bereavement and Grief

Cultural Variations in Bereavement Practices

When it comes to the grieving process, cultural beliefs and practices play a significant role in shaping how individuals and communities mourn the loss of a loved one. Cultural variations in bereavement practices can encompass a wide range of customs, rituals, and traditions that guide the grieving process and provide support to the bereaved. Understanding these cultural variations is essential for healthcare professionals and support workers to provide culturally sensitive and appropriate care during times of loss.

The Significance of Cultural Context

Culture influences our beliefs, values, and behaviors, including how we understand and cope with death. Different cultures have diverse perspectives on death, the afterlife, and the appropriate ways to mourn and honor the deceased. These beliefs and practices are deeply rooted in cultural and religious traditions, which shape the rituals and ceremonies performed during the grieving process.

For example, in some Western cultures, death may be viewed as a private matter, and the focus may be on individual grieving and personal healing. In contrast, in many Asian cultures, death is often seen as a family affair, with an emphasis on collective mourning and honoring the ancestors. In Native American cultures, death is regarded as a part of the life cycle, and the spiritual connection between the living and the deceased is celebrated.

Rituals and Ceremonies

Cultural variations in bereavement practices manifest in the rituals and ceremonies performed after the death of a loved one. These practices provide structure and support to the bereaved and help them navigate the grieving process. Some common cultural variations in bereavement practices include:

- Funerals and memorial services: Funeral practices vary widely across cultures. In some cultures, funerals are elaborate ceremonies involving specific rituals, prayers, and processions. In others, they may be more informal gatherings where family and friends share memories and offer support to the bereaved.

- Mourning attire: Wearing specific clothing or colors as a sign of mourning is a common practice in many cultures. For example, wearing black or white clothing is often seen as a symbol of respect and mourning in Western cultures, while in some Asian cultures, white may be associated with death and is thus avoided.

+ Burial and cremation customs: The way in which the deceased is laid to rest varies across cultures. Some cultures practice burial, while others prefer cremation. There may also be specific rituals and customs associated with the burial or cremation process, such as the use of specific burial sites or the scattering of ashes in meaningful locations.

+ Mourning periods: Different cultures observe mourning periods of varying lengths. These periods may involve certain practices or restrictions, such as refraining from social gatherings, wearing specific clothing, or following dietary restrictions. Mourning periods allow the bereaved to grieve and gradually adjust to life without their loved one.

+ Commemorative rituals: Many cultures have commemorative rituals or ceremonies that are performed at specific intervals after the death. These rituals, such as anniversary memorials or annual remembrance services, provide an opportunity for the bereaved to reflect on their loss and honor the memory of the deceased.

Challenges in Cultural Variations

While cultural variations in bereavement practices are essential for respecting and honoring diverse beliefs and traditions, they can also pose challenges in healthcare and support settings. Some of these challenges include:

+ Communication barriers: Language and cultural differences may hinder effective communication between the bereaved and healthcare professionals or support workers. It is crucial to have interpreters or cultural liaisons available to ensure clear communication and understanding.

+ Clash of beliefs and values: The cultural beliefs and practices of the bereaved may conflict with institutional policies or medical interventions. Understanding and respecting these differences is essential to navigate any potential conflicts and provide person-centered care.

+ Lack of cultural competence: Healthcare professionals and support workers may have limited knowledge or understanding of cultural variations in bereavement practices. It is important to invest in cultural competence training and education to ensure that care is provided in a sensitive and appropriate manner.

Promoting Cultural Sensitivity

To provide culturally sensitive care during times of bereavement, healthcare professionals and support workers can consider the following strategies:

+ Education and awareness: Learning about different cultural beliefs and practices surrounding death and mourning can help healthcare professionals and support workers understand and respect the needs of the bereaved from diverse cultural backgrounds.

+ Open and respectful communication: Creating a safe and non-judgmental environment where the bereaved feel comfortable sharing their cultural beliefs and practices is essential. Active listening and empathy can foster trust and support culturally appropriate care.

+ Collaborative approach: Involving the bereaved and their families in decision-making processes and care planning can ensure that their cultural values and preferences are respected.

+ Partnering with community resources: Collaborating with community organizations or cultural groups can provide valuable insights and support in navigating cultural variations in bereavement practices.

+ Reflective practice: Regularly reflecting on personal biases, assumptions, and cultural beliefs can help healthcare professionals and support workers provide unbiased and culturally sensitive care.

Real-World Example: Cultural Variations in Mourning Practices

To illustrate the impact of cultural variations in bereavement practices, let's consider the example of mourning practices in the African-American community. In this community, funerals are often seen as a celebration of life, with lively music, heartfelt eulogies, and vibrant clothing choices. The bereaved come together for a "homegoing" ceremony, which serves to comfort the family and celebrate the belief in the transition from earthly life to the eternal afterlife.

This cultural variation in bereavement practices highlights the importance of understanding and respecting the traditions and customs of different cultural groups. By recognizing the significance of these practices, healthcare professionals and support workers can provide culturally sensitive care that honors the beliefs and values of the bereaved during their grieving process.

Further Resources

+ Book: "Death and Bereavement Across Cultures" by Colin Murray Parkes, Pittu Laungani, and Bill Young (1997)

+ Article: "Cultural Diversity in Death, Dying, and Grief" by Kenneth J. Doka (2002)

+ Organization: Association for Death Education and Counseling - www.adec.org

Conclusion

Cultural variations in bereavement practices reflect the diverse ways in which different cultures understand and cope with death and loss. Recognizing and respecting these cultural beliefs and practices is essential for healthcare professionals and support workers in providing culturally sensitive care to the bereaved. By understanding the significance of rituals, customs, and traditions, professionals can offer meaningful support during times of grief and mourning.

Ethical Considerations in Supporting Grieving Individuals

Grief is a deeply personal and complex experience that accompanies the loss of a loved one. When individuals are faced with the death of someone significant to them, they often go through a range of emotions, thoughts, and physical sensations. As health and wellness professionals, it is crucial to understand and address the ethical considerations involved in supporting grieving individuals.

The Importance of Compassion and Empathy

Compassion and empathy are the foundations of providing support to grieving individuals. Demonstrating compassion involves recognizing and acknowledging the person's pain, offering comfort, and showing kindness. Empathy, on the other hand, entails understanding and sharing the emotions of the grieving person, without judgment or minimizing their experiences.

Healthcare professionals should strive to cultivate a caring and empathetic attitude in their interactions with grieving individuals. This can be achieved by actively listening to their concerns, validating their feelings, and avoiding platitudes or dismissive statements. It is important to create a safe and nonjudgmental space that allows individuals to express their grief openly.

Respecting Cultural and Individual Differences

Supporting grieving individuals requires a deep respect for cultural and individual differences. Culture plays a significant role in how individuals perceive and express grief. Various cultural practices and rituals may influence the grieving process, such as mourning periods, funeral customs, and religious beliefs about the afterlife.

Health and wellness professionals must be mindful of cultural norms and sensitivities when providing support to diverse populations. This involves understanding and respecting individual and group-specific expressions of grief. By doing so, professionals can provide culturally competent care that aligns with the values and beliefs of the grieving individuals.

Addressing Ethical Challenges Surrounding Self-Care

Healthcare professionals involved in supporting grieving individuals often face ethical challenges related to self-care. Witnessing and engaging with individuals in distress can take an emotional toll on professionals, potentially leading to compassion fatigue or burnout. This can compromise their ability to provide effective support to grieving individuals.

To address these challenges, professionals must prioritize self-care and maintain healthy boundaries. This may include seeking support from colleagues or supervisors, engaging in self-reflection and self-care activities, and accessing professional counseling or therapy when needed. By taking care of their own well-being, professionals can continue to provide compassionate and ethical support to grieving individuals.

Ethical Considerations in Bereavement Support Programs

Bereavement support programs play a crucial role in providing ongoing support to individuals who have experienced the loss of a loved one. These programs are designed to provide emotional, psychological, and practical assistance to individuals throughout their grief journey.

However, ethical considerations arise when designing and implementing bereavement support programs. These considerations include maintaining confidentiality and privacy, respecting the autonomy and agency of the individuals seeking support, and ensuring that the programs are accessible and inclusive.

Health and wellness professionals should prioritize informed consent, ensuring that participants understand and freely choose to engage in the support programs. Additionally, they should establish clear guidelines regarding confidentiality and privacy, assuring individuals that their personal information will be protected.

Inclusivity and accessibility should be key considerations in bereavement support programs. Efforts should be made to accommodate individuals from diverse backgrounds and provide support in a manner that considers their unique needs and preferences.

Unconventional Approach: Art Therapy for Grieving Individuals

An unconventional yet effective approach to supporting grieving individuals is through art therapy. Art therapy utilizes creative processes and materials to help individuals express their emotions, explore their grief, and find healing.

This approach can be particularly useful for individuals who struggle to communicate their feelings verbally or find it challenging to express their grief. Through art therapy, individuals can externalize their emotions, gain insights into their grief experience, and develop coping strategies.

Incorporating art therapy into bereavement support programs allows individuals to engage in a nonverbal and creative process that can facilitate healing and meaning-making. It provides an alternative avenue for expression and can complement traditional talk therapy approaches.

Conclusion

Supporting grieving individuals requires a compassionate and empathetic approach, rooted in a deep understanding of cultural and individual differences. Health and wellness professionals must navigate ethical challenges surrounding self-care and prioritize the creation of inclusive and accessible bereavement support programs. By incorporating unconventional approaches like art therapy, professionals can provide meaningful and effective support to individuals navigating the complex journey of grief.

End-of-life Care for Families and Caregivers

End-of-life care is a crucial aspect of the healthcare system, ensuring that individuals nearing the end of their lives receive appropriate support and comfort. However, it is not just the patients who require care during this time; their families and caregivers also need support and guidance. This section explores the ethical considerations involved in providing end-of-life care for families and caregivers.

Importance of Supporting Families and Caregivers

When a loved one is nearing the end of their life, it can be an emotionally challenging and overwhelming experience for families and caregivers. They often experience a range of emotions, including grief, anxiety, guilt, and sadness. It is essential for healthcare professionals to acknowledge and address these emotions, providing support and guidance to help families and caregivers navigate the complexities of end-of-life care.

Challenges Faced by Families and Caregivers

Families and caregivers face various challenges when providing care to individuals nearing the end of their lives. These challenges may include:

- Emotional distress: Families and caregivers may experience significant emotional distress due to witnessing their loved one's suffering and knowing that their time together is limited.

- Physical exhaustion: Providing care for a terminally ill individual can be physically demanding and exhausting, leading to burnout and fatigue.

- Decision-making: Families and caregivers often have to make difficult decisions on behalf of their loved ones, such as choosing between different treatment options or deciding when to initiate palliative care.

- Communication: Effective communication between healthcare professionals, families, and caregivers is critical during end-of-life care. However, it can be challenging to navigate discussions about prognosis, treatment options, and end-of-life preferences.

- Financial burden: End-of-life care can be expensive, placing a significant financial burden on families and caregivers. This burden can create additional stress and anxiety during an already challenging time.

Ethical Considerations

In providing support and care for families and caregivers, several ethical considerations come into play. These include:

Respecting Autonomy and Decision-Making   Families and caregivers should be actively involved in the decision-making process regarding the care of their loved ones.   Healthcare professionals should respect their autonomy by providing comprehensive information about treatment options, prognosis, and end-of-life care, allowing them to make informed decisions that align with their values and wishes.

Maintaining Confidentiality and Privacy   Respecting the privacy and confidentiality of families and caregivers is crucial during end-of-life care. Healthcare professionals should ensure that sensitive information is securely protected, and that only authorized individuals have access to the patient's medical records.   This not only maintains the privacy of families and caregivers but also fosters trust and openness in the healthcare relationship.

Providing Emotional Support   Families and caregivers require emotional support throughout the end-of-life care journey. Healthcare professionals should create a compassionate and supportive environment, offering opportunities for families and caregivers to express their emotions and concerns. This can include providing access to counseling services, support groups, and bereavement resources.

Addressing Cultural and Spiritual Needs   Cultural and spiritual beliefs play a significant role in end-of-life care. Healthcare professionals should be sensitive to the diverse cultural and spiritual backgrounds of families and caregivers, recognizing and respecting their beliefs and practices.   This may involve collaborating with religious or cultural advisors and ensuring that religious and cultural rituals are honored.

Providing Respite Care   Given the physical and emotional demands of providing end-of-life care, families and caregivers may benefit from respite care. This involves temporary relief from caregiving duties, allowing families and caregivers to recharge and take care of their own well-being. Healthcare professionals should recognize the importance of respite care and help families and caregivers identify and access appropriate support services.

Facilitating Bereavement Support   The end of a patient's life marks the beginning of the bereavement process for families and caregivers.   Healthcare professionals should provide bereavement support, connecting families and caregivers with resources and counseling services to help them cope with their loss.

This can include support groups, individual therapy, and educational materials on the grieving process.

Case Study: Supporting Families and Caregivers

A family is caring for their elderly mother, who has been diagnosed with a terminal illness. The burden of caregiving has taken a toll on the family's emotional well-being, and they are struggling to make decisions regarding her end-of-life care. The healthcare team recognizes the importance of supporting the family and takes the following steps:

+ The healthcare team schedules a family meeting to discuss the mother's prognosis, treatment options, and goals of care. They provide a comprehensive explanation of the available resources for end-of-life care.

+ The team acknowledges the emotional distress experienced by the family members and offers them access to counseling services to help them cope with their feelings of sadness, guilt, and anxiety.

+ The healthcare team connects the family with a social worker who can assist them in navigating the financial aspects of end-of-life care, providing information about available financial resources and support.

+ Cultural considerations are taken into account, and the team collaborates with the family's religious leader to ensure that cultural and spiritual practices are respected and integrated into the end-of-life care plan.

+ The team discusses the importance of respite care with the family and helps them identify respite care options available to temporarily relieve their caregiving responsibilities.

+ As the patient nears the end of her life, the healthcare team continues to offer bereavement support, providing the family with information on support groups and counseling services to assist them in navigating the grieving process.

Conclusion

End-of-life care for families and caregivers is a complex and sensitive area that requires healthcare professionals to navigate various ethical considerations. By respecting autonomy, providing emotional support, addressing cultural and spiritual needs, and offering respite care and bereavement support, healthcare professionals can help families and caregivers through this challenging time, ensuring they receive the care and support they need.

Ethical Issues in Public Health Emergencies

The Ethics of Crisis Standards of Care

Allocating Scarce Resources During a Public Health Emergency

During a public health emergency, such as a pandemic or natural disaster, the demand for healthcare resources often exceeds the available supply. This situation poses ethical challenges in allocating scarce resources in a fair and equitable manner. In this section, we will explore the ethical considerations involved in making allocation decisions during such emergencies.

The Importance of Fair Allocation

Fair allocation of scarce resources is crucial to ensuring that the benefits and burdens of healthcare are distributed justly. In times of crisis, the goal is to maximize the overall well-being of the population and minimize harm. However, certain ethical principles and theories can guide decision-making in resource allocation.

Utilitarianism

Utilitarianism, a consequentialist ethical theory, suggests that the action that produces the greatest overall happiness or well-being for the greatest number of people is ethically correct. Applied to resource allocation, utilitarianism would prioritize allocating resources to those individuals who are most likely to benefit and survive.

Principle of Need

The principle of need, derived from the principle of distributive justice, asserts that resources should be allocated based on individuals' needs. In a public health emergency, this principle would prioritize allocating resources to individuals with the greatest urgency and potential for survival.

Proportionality

The principle of proportionality suggests that resources should be allocated in proportion to the severity of illness or injury. In other words, those who are sicker or at higher risk should receive a greater share of resources. This principle aims to achieve fairness in resource allocation.

Prioritization Criteria

To allocate scarce resources during a public health emergency, specific criteria must be established to guide decision-making. These criteria should be transparent, consistent, and impartial. Several factors can be considered when determining the priority of resource allocation:

Medical Criteria

Medical criteria include the severity of illness, prognosis, and the likelihood of benefiting from the resource. Patients who are critically ill, have a high chance of survival, or will experience significant improvement with treatment may receive higher priority.

Age Considerations

Age is often seen as a relevant factor, as older individuals may have a lower life expectancy and fewer remaining years of life. However, age alone should not be the sole criterion for allocating resources, as it may lead to discrimination against older adults.

Ethical Considerations

Ethical considerations encompass principles such as equity, fairness, and non-discrimination. It is important to consider the needs and interests of marginalized and vulnerable populations, who may have historically faced disparities in access to healthcare.

Additional Factors

Other factors, such as the role of healthcare workers in the response efforts, community impact, and public health considerations, may also be taken into account during resource allocation decisions.

Challenges and Solutions

Allocating scarce resources in a public health emergency poses several challenges:

Limited Resources

The insufficient availability of essential resources like ventilators, personal protective equipment (PPE), or ICU beds can lead to difficult decisions about who receives access. To address this challenge, strategies like optimizing resource allocation through triage systems and alternative care settings can be employed.

Allocation Bias

The potential for bias in allocation decisions is a concern. Implicit biases based on race, ethnicity, gender, socioeconomic status, or disability must be identified and minimized through structured decision-making processes and oversight.

Transparency and Accountability

Transparency in decision-making processes is crucial to maintaining public trust and ensuring fairness. Clear guidelines and criteria should be communicated to the public, and mechanisms for oversight and appeals should be established.

Legal and Policy Frameworks

It is essential to have legal and policy frameworks in place that outline the ethical principles, guidelines, and procedures for resource allocation during emergencies. These frameworks can provide a framework for decision-making and help resolve disputes.

Real-World Examples

During the COVID-19 pandemic, healthcare systems around the world faced the challenge of allocating scarce resources. In Italy, for instance, hospitals developed guidelines for triaging patients based on clinical criteria, giving priority to those with the highest chance of survival.

In the United States, some states and hospitals implemented crisis standards of care, which establish guidelines for allocating resources during emergencies. These standards outline the prioritization criteria and decision-making processes to ensure fair and equitable allocation.

Conclusion

Allocating scarce resources during a public health emergency is a complex ethical endeavor. It requires balancing the principles of utilitarianism, need, and proportionality while addressing challenges such as limited resources, bias, transparency, and accountability. By establishing clear criteria, involving diverse perspectives, and fostering public trust, we can strive for fair and equitable allocation to promote the well-being of all individuals during times of crisis.

Triage and Fairness in Crisis Standards of Care

During public health emergencies, such as pandemics or natural disasters, healthcare systems may face an overwhelming surge of patients needing immediate medical attention. The allocation of scarce resources becomes a significant concern, and crisis standards of care (CSC) are often implemented to guide decision-making and prioritize patients based on their likelihood of survival and potential for benefiting from limited resources. Triage plays a crucial role in these standards by determining the order in which patients receive medical treatment.

Background

Triage is a process used to evaluate and prioritize patients based on the severity of their condition and the resources available. The term "triage" originated from French military medicine during the Napoleonic Wars, where it was used to sort wounded soldiers based on the urgency of their injuries. In modern healthcare, triage is commonly employed in emergency departments to allocate care based on the urgency and severity of patients' conditions.

In crisis situations, triage becomes even more critical as the need for medical resources exceeds their availability. Crisis standards of care provide guidelines for healthcare providers to make ethical and fair decisions regarding the allocation of scarce resources during extraordinary circumstances.

Principles of Triage in Crisis Standards of Care

When implementing triage in crisis standards of care, several ethical principles guide decision-making to ensure fairness and justice:

+ **Maximize Benefits:** The primary goal of triage is to save as many lives as possible and minimize overall harm. Patients with a higher likelihood of survival and greater potential to benefit from treatment are given priority access to limited resources. This principle aims to maximize the overall health benefit for the population.

+ **Fairness and Equity:** Triage decisions must be fair and equitable. Factors such as age, race, gender, or social status should not influence resource allocation. Instead, decisions should be based solely on medical need, urgency, and the potential to benefit from the available resources.

+ **Proportionality:** The allocation of limited resources should be proportional to patients' clinical needs. Those with more severe conditions or a greater chance of survival with treatment should receive priority over patients with less urgent or severe conditions. This principle aims to ensure that resources are allocated in a manner that aligns with medical necessity.

+ **Transparency and Accountability:** Triage decisions should be transparent, clearly communicated, and accountable. Healthcare providers should provide justifications for their decisions, ensure clear documentation, and regularly review and update triage protocols as necessary. Transparency increases public trust in the decision-making process and promotes accountability among healthcare professionals.

Ethical Challenges in Triage Decision-making

While the principles mentioned above provide a framework for fair and just triage, several ethical challenges can arise in the implementation of crisis standards of care:

+ **Resource Scarcity:** The scarcity of resources during a crisis adds an ethical dimension to triage decision-making. Healthcare providers may face difficult choices when there are not enough resources to treat all patients adequately. This scarcity introduces ethical dilemmas, such as deciding who receives potentially life-saving treatments and who does not.

+ **Subjectivity and Bias:** Triage decision-making can be influenced by subjectivity and unconscious biases, which may lead to inequitable resource allocation. Healthcare providers must be aware of their biases and strive to minimize their impact on triage decisions. Training and clear guidelines can help mitigate the influence of bias and ensure more objective decision-making.

+ **Emotional and Psychological Toll:** Triage decisions can be emotionally and psychologically challenging for healthcare providers. The responsibility of determining who receives potentially life-saving treatments and who does not can cause moral distress and distressing emotions. Providing adequate support and resources for healthcare providers' well-being is essential.

Addressing Ethical Challenges

To address the ethical challenges associated with triage decision-making in crisis standards of care, various strategies can be employed:

+ **Clear Guidelines and Protocols:** Developing clear and evidence-based triage guidelines and protocols can help healthcare providers make consistent and fair decisions. These guidelines should consider the principles of triage and crisis standards of care, ensuring transparency and accountability in decision-making processes.

+ **Education and Training:** Healthcare providers should receive training on ethical decision-making, including triage protocols, bias recognition and mitigation, and communication strategies. Training can help ensure that triage decisions are made based on medical need and are not influenced by personal biases, perceptions, or prejudices.

+ **Engaging Stakeholders:** Engaging diverse stakeholders, including healthcare professionals, ethicists, community representatives, and public health experts, in the development of triage protocols is crucial. This inclusive approach helps ensure that various perspectives and concerns are considered and that decisions are accepted and understood by the broader community.

+ **Continuous Evaluation and Improvement:** Regular evaluation and improvement of triage protocols are necessary to address emerging ethical challenges and incorporate lessons learned from previous crises. Flexibility in adapting protocols to changing circumstances and feedback from healthcare providers and the community enhance the fairness, effectiveness, and ethical soundness of triage decision-making.

Case Study: Triage during a Pandemic

To illustrate the complexities of triage decision-making, let's consider a hypothetical case during a pandemic. In this scenario, a hospital's intensive care unit (ICU) has reached maximum capacity, with more critically ill patients requiring ICU-level care than the hospital can accommodate.

The triage team evaluates each patient using established triage criteria, including physiological status, comorbidities, and overall prognosis. Based on these criteria, patients are categorized into three groups: those who are likely to survive even without ICU care, those who will likely benefit from ICU care, and those with a low chance of survival despite ICU care.

Using the principle of maximizing benefits, patients in the second group, who have the highest chances of survival with ICU care, are prioritized. However, the decision is not solely based on survival probability. The principles of fairness and equity ensure that age, race, gender, or social status do not influence the decision-making process.

Continuous evaluation and adjustment of triage protocols based on emerging evidence and real-time data can help improve the fairness and effectiveness of these decisions. Regular feedback and engagement with the community and other stakeholders can increase public trust and promote a sense of shared responsibility during the crisis.

Conclusion

Triage and fairness in crisis standards of care play a vital role in allocating scarce resources during public health emergencies. Adhering to ethical principles, such as maximizing benefits, fairness and equity, proportionality, transparency, and accountability, helps healthcare providers make just decisions in challenging circumstances.

Recognizing and addressing the ethical challenges that arise in triage decision-making, such as resource scarcity, subjectivity and bias, and the emotional toll on healthcare providers, are crucial steps in promoting ethical and just triage practices.

By continuously evaluating and improving triage protocols, engaging stakeholders, and providing education and support for healthcare professionals, we can strive to ensure that the allocation of limited resources during crises is based on ethical principles and promotes the well-being of individuals and communities alike.

Legal and Policy Considerations in Public Health Emergencies

Public health emergencies, such as pandemics or natural disasters, pose unique challenges that require careful consideration of legal and policy frameworks. In these situations, swift action is necessary to protect the health and well-being of the population, but it must also be balanced with respect for individual rights and the principles of justice. This section examines the ethical dilemmas and legal considerations that arise during public health emergencies and explores potential solutions to address them.

Emergency Powers and Legal Authority

In the face of a public health emergency, governments may invoke emergency powers and legal authority to implement measures that would not be applicable under normal circumstances. These measures can include quarantine orders, travel restrictions, compulsory vaccination, or the allocation of scarce resources. However, the exercise of emergency powers must be justified, transparent, and proportionate to the threat at hand.

One of the key challenges is determining the extent to which individual rights can be temporarily limited for the greater good. Balancing public health and individual rights is a delicate matter, and the use of emergency powers should be guided by principles of necessity, proportionality, and legality. Legal frameworks should outline clear criteria for the application and termination of emergency powers and establish mechanisms for independent oversight and accountability.

Equity and Non-Discrimination

Public health emergencies have the potential to exacerbate existing health disparities and deepen social inequalities. Vulnerable populations, such as low-income communities, racial and ethnic minorities, and individuals with disabilities, are often disproportionately affected. It is crucial to ensure that legal and policy responses to public health emergencies do not further marginalize these groups.

Efforts should be made to prioritize equitable access to healthcare and essential resources, including testing, treatment, and vaccines. Non-discrimination principles should guide decision-making processes to mitigate the impact on vulnerable populations. It is important to consider cultural and linguistic diversity, as well as the specific needs of individuals with disabilities or limited access to healthcare.

Transparency and Communication

Transparent and timely communication is essential during public health emergencies. Accurate information should be readily available to the public to promote trust, reduce misinformation, and facilitate cooperation. Legal frameworks should mandate regular updates and provide guidelines on disseminating information in a clear and accessible manner.

Effective communication involves not only relaying information but also actively engaging the public in decision-making processes. Soliciting input from community leaders, healthcare professionals, and affected individuals can help identify concerns, improve understanding, and promote ownership of response strategies. Inclusivity should be a guiding principle to ensure that diverse perspectives are considered.

International Cooperation and Coordination

Public health emergencies often transcend national boundaries, requiring international cooperation and coordination. Legal frameworks should facilitate collaboration among countries and international organizations to ensure a unified response. This collaboration includes sharing information, resources, and best practices, as well as coordinating research efforts.

However, international cooperation in the context of public health emergencies also raises ethical questions. Issues related to data sharing, intellectual property rights, and equitable distribution of resources need to be addressed. Policymakers must strive to strike a balance between global solidarity and national interests, with a focus on promoting the common good and supporting the most vulnerable countries.

Building Resilience and Preparedness

To effectively respond to public health emergencies, legal and policy frameworks should prioritize building resilience and preparedness in advance. This includes establishing robust surveillance systems, developing emergency response plans, and investing in public health infrastructure. By proactively addressing these issues, governments can mitigate the impact of emergencies and improve the overall health and well-being of their populations.

Legal frameworks should outline responsibilities and resources needed to support preparedness efforts. They should also ensure that lessons learned from previous emergencies are incorporated into future planning. Ethical considerations, such as equitable distribution of resources and a focus on vulnerable populations, should be integrated into preparedness strategies.

Conclusion

Legal and policy considerations play a crucial role in shaping the response to public health emergencies. By upholding principles of equity, transparency, and international cooperation, governments can effectively address the challenges posed by emergencies while respecting individual rights. Building resilience and preparedness in advance is critical to minimize the impact of future emergencies. By adopting a comprehensive and ethical approach, societies can navigate public health emergencies with compassion, justice, and solidarity.

Public Health Surveillance and Privacy

Balancing Public Health Interests and Individual Privacy Rights

In the context of public health emergencies, there is often a tension between protecting public health interests and respecting individual privacy rights. On one hand, governments and public health authorities need access to timely and accurate information in order to effectively respond to a crisis and prevent the spread of disease. On the other hand, individuals have the right to privacy and may be concerned about the collection and use of their personal health information.

The Importance of Public Health Interests

Public health is focused on promoting and protecting the health of entire populations. During a public health emergency, such as a pandemic, it is crucial to have access to timely and accurate information in order to make informed decisions and take appropriate actions to mitigate the impact of the crisis. This includes identifying sources of infection, implementing control measures, and monitoring the effectiveness of interventions.

In the case of infectious diseases, such as COVID-19, contact tracing plays a crucial role in identifying individuals who may have been exposed to the virus. By identifying and isolating these individuals, the spread of the disease can be effectively controlled. This requires collecting and sharing personal health information, such as test results and contact details, to enable public health authorities to identify and notify potentially infected individuals.

Respecting Individual Privacy Rights

Respecting individual privacy rights is a fundamental ethical principle that ensures individuals have control over their personal information and are protected from unauthorized access or use. Privacy is essential for maintaining autonomy and dignity, as well as fostering trust in healthcare systems and public health interventions.

In the context of public health emergencies, individuals may have concerns about the collection and use of their personal health information. They may worry about the potential misuse of their data, such as discrimination, stigmatization, or breaches of confidentiality. These concerns can impact individuals' willingness to comply with public health measures and may hinder the effectiveness of public health interventions.

Balancing Public Health Interests and Individual Privacy Rights

Balancing public health interests and individual privacy rights requires a delicate and thoughtful approach. It is important to find a middle ground that allows for effective public health responses while respecting individual privacy and maintaining public trust. Several key considerations can guide the balancing act:

1. **Transparency and Consent:** Transparent communication about the purpose, scope, and limitations of data collection is essential. Individuals should be informed about how their personal health information will be used and have the opportunity to provide informed consent. Clear consent processes empower individuals to make informed decisions about sharing their personal information.

2. **Data Minimization and Anonymization:** Collecting only necessary data and minimizing the collection of personally identifiable information helps reduce privacy risks. Data should be anonymized whenever possible to ensure that individuals cannot be personally identified. Aggregated data can still provide valuable insights for public health purposes while protecting individual privacy.

3. **Security and Confidentiality:** Implementing robust security measures to protect personal health information is critical. Encryption, access controls, and secure data storage can help prevent unauthorized access and maintain confidentiality. Public health authorities should be held accountable for ensuring the security and privacy of collected data.

4. **Data Purpose and Retention:** Data should only be used for the specific purpose for which it was collected and should not be retained for longer than necessary. When the public health emergency ends, data should be securely deleted or anonymized to minimize the risk of unintended consequences or misuse.

5. **Ethical Oversight and Regulation:** Establishing ethical frameworks and clear regulations for data collection, use, and sharing is crucial. Independent ethical review boards and regulatory agencies can help ensure that public health measures are carried out in a manner that respects individual privacy rights and aligns with ethical principles.

6. **Public Education and Engagement:** Empowering individuals through public education and engagement can foster understanding and trust. Clear communication about the importance of data collection, the measures in place to protect privacy, and the benefits to public health can help address concerns and encourage participation.

Real-World Example: Contact Tracing Apps

A real-world example of balancing public health interests and individual privacy rights is the use of contact tracing apps during the COVID-19 pandemic. These apps have been developed to assist in identifying and notifying individuals who may have come into close contact with someone infected with the virus.

To ensure privacy protection, many contact tracing apps have implemented privacy-enhancing measures. For example, some apps use Bluetooth technology to anonymously exchange information between devices, without revealing personal details. User consent is typically required before using the app, and users have the ability to disable or delete the app at any time. Additionally, data collected is often subject to strict retention limits, and efforts are made to ensure that it is used only for contact tracing purposes.

By implementing these privacy-centric measures and engaging in transparent communication, contact tracing apps aim to strike a balance between public health interests and individual privacy rights. Such examples demonstrate that it is possible to leverage technology to enhance public health responses while respecting individual privacy.

Conclusion

Balancing public health interests and individual privacy rights is a complex endeavor, particularly in the context of public health emergencies. However, with transparency, consent, data minimization, secure practices, and public education, it is possible to find a middle ground that protects public health and respects individual privacy. Ethical oversight and regulation play a crucial role in ensuring that these considerations are upheld, and public trust is maintained. Striving for this balance is essential for navigating moral dilemmas, protecting justice, and shaping a more equitable, just, and sustainable world of health and wellness.

Health Data Collection and Use in Public Health Emergencies

During public health emergencies, such as pandemics or natural disasters, the collection and use of health data become crucial for effective response and decision-making. Ethical considerations surrounding the collection and use of health data in these situations are of paramount importance to safeguard individual privacy, ensure informed consent, and maintain public trust. This section will explore the ethical challenges and principles involved in health data collection and use during public health emergencies.

The Importance of Health Data in Public Health Emergencies

Health data plays a pivotal role in understanding the spread of diseases, identifying vulnerable populations, and implementing effective interventions during public health emergencies. Timely and accurate data collection allows public health authorities to track disease transmission, assess the impact on the community, and allocate resources strategically.

In the context of public health emergencies, health data can be collected through various sources, including surveillance systems, electronic health records, mobile applications, and wearable devices. These data may include information about symptoms, test results, contact tracing, and demographic details. The collection, storage, and use of these data raise ethical concerns that need careful consideration.

Ethical Principles in Health Data Collection

When collecting health data during a public health emergency, several ethical principles should guide the process. These principles include:

1. **Privacy and Confidentiality:** It is essential to protect the privacy and confidentiality of individuals whose data is collected. Health data should be de-identified whenever possible, and strict security measures should be in place to prevent unauthorized access.

2. **Informed Consent:** Informed consent is a fundamental ethical principle that ensures individuals are fully informed about the purpose, risks, and benefits of data collection. During emergencies, the requirement for informed consent may be modified while still respecting the autonomy and rights of individuals.

3. **Necessity and Proportionality:** The collection and use of health data should be necessary and proportionate to the public health emergency at hand. Only data required for preventing, controlling, and mitigating the emergency should be collected.

4. **Transparency and Accountability:** Public health authorities should be transparent about the purposes and processes of health data collection. Accountability mechanisms must be in place to ensure responsible data handling and prevent misuse.

Ethical Challenges in Health Data Use

While the collection of health data during public health emergencies is crucial, its use poses ethical challenges that must be addressed. Some key challenges include:

1. **Data Sharing and Interoperability:** Ensuring interoperability and secure sharing of health data between different agencies and institutions is essential for a coordinated emergency response. However, data sharing must adhere to privacy and confidentiality standards to protect individuals' rights and prevent misuse.

2. **Data Quality and Accuracy:** During emergencies, there is often a need for rapid data collection and analysis. This urgency can compromise data quality and accuracy. It is critical to maintain data integrity and minimize errors to ensure effective decision-making.

3. **Stigmatization and Discrimination:** Health data, especially when linked to specific geographic areas or demographics, can reinforce stereotypes, stigmatize communities, and perpetuate discrimination. The ethical use of health data should actively mitigate these risks to protect vulnerable populations.

4. **Surveillance and Privacy Trade-offs:** Public health emergencies may warrant expanded surveillance measures to track the spread of diseases. However, increased surveillance can infringe upon privacy rights. Balancing the need for public health protection with the preservation of individual privacy is a delicate ethical challenge.

Examples and Real-World Application

An example of health data collection and use during a public health emergency is contact tracing for COVID-19. Mobile applications have been developed to track individuals' interactions and potential exposure to the virus. While contact tracing can be effective in controlling the spread of the disease, it raises concerns about data security, consent, and potential stigmatization. Ethical guidelines and technical safeguards are essential to address these challenges and protect individuals' rights.

To ensure ethical data collection and use, public health authorities can establish data governance frameworks that include measures such as data anonymization, robust security protocols, and strict access controls. Additionally, public engagement and transparent communication can help build trust and address concerns regarding health data privacy and confidentiality.

Summary

Health data collection and use in public health emergencies present ethical challenges that necessitate careful scrutiny and adherence to fundamental principles. Protecting privacy and confidentiality, ensuring informed consent, and maintaining data quality are crucial considerations. Balancing the need for data-driven decision-making with safeguarding individual rights and public trust is essential for effective emergency response. By upholding ethical standards, we can harness the potential of health data to mitigate the impact of public health emergencies and promote the well-being of individuals and communities.

Strategies for Building Public Trust in Surveillance Measures

In the face of public health emergencies and the need for effective surveillance measures, it is crucial to build public trust in order to ensure the successful implementation and acceptance of these measures. Trust plays a key role in facilitating cooperation, compliance, and overall public health outcomes. However, surveillance measures, particularly those involving the collection and use of personal health data, can raise concerns regarding privacy, security, and the potential for abuse. In this section, we will explore strategies for building public trust in surveillance measures, focusing on transparency, accountability, and public engagement.

Transparency in Surveillance Practices

Transparency is vital in addressing public concerns related to surveillance measures. Openly sharing information about the purpose, scope, and methods of surveillance can help alleviate fears and build trust. Here are some strategies for promoting transparency:

1. Clear Communication: Provide clear and concise explanations of the need for surveillance measures, the data being collected, and how it will be used.

2. Plain Language: Avoid technical jargon and use plain language to ensure information is easily understood by all members of the public.

3. Public Consultation: Seek input from the public and relevant stakeholders in the development and implementation of surveillance measures. This involvement fosters a sense of ownership and can help identify and address potential concerns.

4. Public Reporting: Regularly provide updates on surveillance activities, including the results of data analyses and any actions taken as a result of the findings. This demonstrates accountability and reinforces transparency.

Accountability in Surveillance Practices

Accountability is crucial in building trust and ensuring that surveillance measures are used responsibly. When individuals feel that their data is being handled ethically and that appropriate safeguards are in place, they are more likely to trust the system. Here are some strategies for promoting accountability:

1. Legal Frameworks: Establish clear legal frameworks to regulate the collection, storage, and use of personal health data. These frameworks should include provisions for obtaining informed consent, ensuring data security, and addressing potential breaches.

2. Independent Oversight: Create independent oversight bodies to monitor and evaluate surveillance practices. These bodies should have the authority to investigate complaints, assess compliance with legal requirements, and recommend improvements.

3. Strong Security Measures: Implement robust security measures to protect personal health data from unauthorized access, use, or disclosure. This includes encryption, access controls, and regular audits of security protocols.

4. Data Minimization: Adopt a principle of data minimization, collecting only the necessary data for surveillance purposes. Clear guidelines should be provided

to minimize the risk of collecting and retaining excessive or unnecessary personal information.

Public Engagement in Surveillance Practices

Public engagement is essential for building trust and ensuring that surveillance measures align with societal values and expectations. By involving the public in decision-making processes, their concerns and perspectives can be taken into account, fostering a sense of ownership and legitimacy. Here are some strategies for promoting public engagement:

1. Education and Awareness Campaigns: Conduct education and awareness campaigns to help the public understand the importance and benefits of surveillance measures. Address common misconceptions and provide accurate information to dispel fears.

2. Stakeholder Involvement: Engage with community leaders, advocacy groups, and other stakeholders to understand their concerns and gather input on surveillance practices. This collaborative approach helps to address diverse perspectives and build consensus.

3. Privacy Enhancing Technologies: Explore the use of privacy-enhancing technologies, such as differential privacy and secure multiparty computation, to protect individual privacy while still enabling effective data analysis for surveillance purposes.

4. Feedback Mechanisms: Establish mechanisms for individuals to provide feedback, ask questions, and voice concerns about surveillance practices. Actively address these concerns and provide timely and meaningful responses to build trust.

It is important to note that building public trust in surveillance measures is an ongoing process. As technology evolves and new challenges arise, strategies for transparency, accountability, and public engagement will need to adapt. By continually engaging with the public, monitoring emerging issues, and refining surveillance practices, we can foster a culture of trust and ensure the effective and ethical use of surveillance measures for public health purposes.

Public Health Communication and Ethics

Accuracy and Transparency in Public Health Communication

Public health communication plays a crucial role in promoting the well-being and safety of individuals and communities. It serves as a vehicle for disseminating accurate information, raising awareness about health risks, and encouraging behavior change. Accuracy and transparency are essential principles that guide effective public health communication. In this section, we will explore the importance of accuracy and transparency in public health messages, the challenges associated with achieving them, and strategies for promoting these principles in communication.

The Significance of Accuracy in Public Health Communication

Accurate information is the foundation of public health communication. It enables individuals to make informed decisions about their health and well-being. Inaccurate or misleading information can have severe consequences, leading to misunderstandings, poor health outcomes, and even public panic. Therefore, it is crucial for public health messages to be based on scientific evidence, verified data, and authoritative sources.

Achieving accuracy in public health communication can be challenging due to various reasons. One common challenge is the fast-paced nature of emerging health issues. During public health emergencies, such as a pandemic or an outbreak, information is rapidly evolving, and uncertainty is high. Public health experts and communicators must navigate this complexity while providing timely and accurate information to the public.

Another challenge is the dissemination of misinformation through various channels, such as social media platforms. Misinformation can spread quickly, leading to confusion and mistrust among the public. Public health communicators need to actively debunk myths, correct false information, and ensure that accurate information is accessible and easily understood.

Transparency in Public Health Communication

Transparency is another essential aspect of effective public health communication. It involves openness, honesty, and clear communication about the sources of information, the decision-making processes, and the limitations of knowledge. Transparency builds trust, promotes engagement, and empowers individuals to make informed choices.

Transparency is particularly critical during public health emergencies when decisions are made rapidly, and there is often limited information available. Public health communicators must be transparent about the uncertainties, potential risks, and ongoing research. By openly discussing the limitations and evolving nature of the knowledge, they can manage public expectations and maintain trust.

However, achieving transparency in public health communication can be challenging. There may be tensions between the need to provide timely information and the need for thorough verification and analysis. Balancing transparency with the potential for causing panic or confusion requires careful judgment and expertise.

Strategies for Promoting Accuracy and Transparency

Promoting accuracy and transparency in public health communication requires intentional efforts and strategies. Here are some key strategies that can be employed:

1. **Clear and Accessible Language:** Public health messages should use plain language that is easy for the general public to understand. Technical jargon and complex terminology can create confusion and hinder comprehension. Clear and accessible language ensures that accurate information reaches a broader audience.

2. **Utilizing Multiple Communication Channels:** Public health communicators should utilize a variety of communication channels to reach different segments of the population. This includes traditional media, social media platforms, websites, mobile applications, and community-based organizations. Using diverse channels allows for wider dissemination of accurate information.

3. **Engaging with Communities:** Engaging with communities is crucial for promoting accuracy and transparency. Public health communicators should actively involve community leaders, influencers, and trusted spokespersons in the dissemination of information. This helps build trust, addresses specific community concerns, and ensures that the information is culturally appropriate.

4. **Collaboration with Experts:** Public health agencies should collaborate with experts, such as scientists, epidemiologists, and healthcare professionals, to ensure the accuracy of information. Consulting subject matter experts helps validate and verify the information before it is shared with the public.

5. **Fact-checking and Correcting Misinformation:** Public health communicators have a responsibility to fact-check information, especially during public health emergencies when misinformation is rampant. Timely identification and correction of false information are essential to counter the spread of inaccuracies.

6. **Ongoing Evaluation and Improvement:** Continuous evaluation of public health communication efforts is necessary to identify areas for improvement. Feedback from the public and stakeholders can provide valuable insights into the effectiveness of communication strategies and help refine them for better accuracy and transparency.

Case Study: Accuracy and Transparency in COVID-19 Communication

The COVID-19 pandemic presented significant challenges in public health communication, highlighting the importance of accuracy and transparency. Governments, health organizations, and public health agencies faced the task of conveying rapidly changing information about a novel virus while addressing public concerns.

Transparency was critical in providing updates on the evolving nature of the virus, the effectiveness of preventive measures, and the development of vaccines. Public health agencies held regular press briefings, provided updates on case numbers and testing, and shared data on vaccine efficacy and safety. By being transparent about the uncertainties and ongoing research, they maintained public trust and confidence.

At the same time, ensuring accuracy was challenging due to the vast volume of information circulating, including misinformation and conspiracy theories. Public health agencies partnered with fact-checking organizations to debunk myths and provide accurate information through various channels, including social media platforms. They engaged with influencers and celebrities to amplify accurate messages and combat misinformation.

The COVID-19 pandemic demonstrated the importance of accurate and transparent communication in saving lives, reducing anxiety, and promoting preventive behaviors. It underscored the need for ongoing evaluation and improvement of communication strategies to address emerging challenges effectively.

Conclusion

Accuracy and transparency are fundamental principles in public health communication. Providing accurate information ensures that individuals can make informed decisions about their health and well-being. Transparency builds trust and empowers individuals to engage in healthier behaviors.

While challenges exist, strategies such as clear language, utilizing multiple communication channels, engaging with communities, collaborating with experts, fact-checking, and ongoing evaluation can promote accuracy and transparency. The case of COVID-19 highlighted the significance of these principles in navigating public health emergencies.

By prioritizing accuracy and transparency, public health communicators can foster an informed and engaged society, ultimately contributing to better health outcomes for individuals and communities.

Balancing Individual Rights and Public Health in Communication

In the context of public health emergencies, effective communication plays a crucial role in protecting the well-being of individuals and the broader community. However, ethical dilemmas often arise when balancing individual rights to privacy and freedom of information with the need to safeguard public health. This section will explore the challenges and considerations in striking a balance between these competing interests.

The Importance of Individual Rights

Respecting individual rights is a fundamental principle in ethical decision-making. Privacy rights, freedom of speech, and autonomy are foundational principles in democratic societies. In the realm of public health communication, individuals have the right to control their personal health information and make informed choices about their health. These rights are protected by laws and regulations that prioritize individual autonomy and privacy.

The Need for Public Health Protection

At the same time, public health emergencies require strategies that prioritize the protection of the larger community. Communicating accurate and timely information is critical for preventing the spread of contagious diseases, minimizing harm, and saving lives. In situations where individual actions have implications for public health, striking a balance between individual rights and public health protection becomes essential.

Ethical Considerations in Balancing Individual Rights and Public Health

1. Privacy protection: Public health agencies and officials must navigate the delicate balance between disclosing necessary information to the public and protecting individuals' privacy. Striking a balance involves de-identifying data

whenever possible, using aggregated data whenever feasible, and ensuring that any personal information disclosed is strictly necessary for public health purposes.

2. Honesty and transparency: Open and transparent communication builds trust between individuals and public health authorities. During a public health emergency, it is crucial to provide accurate and timely information to the public, including potential risks, preventive measures, and available resources. Balancing the duty to inform the public with the responsibility to avoid unnecessary panic or fear is an ongoing challenge.

3. Tailored messaging: Public health communication should consider the diverse needs and circumstances of different individuals and communities. This requires using language and channels that are accessible and culturally appropriate. Balancing individual information needs with the broader public's need for comprehensive guidance is an ongoing ethical consideration.

4. Consent and opt-out options: In situations where public health measures may involve sharing personal information or participation in contact tracing, individuals should have the option to provide informed consent or opt-out. It is essential to respect individuals' autonomy and privacy while still implementing effective public health strategies.

5. Public health campaigns: Balancing individual rights with public health goals in communication campaigns requires thoughtful considerations. Strategies that aim to influence behavior change should respect individuals' autonomy, avoid coercion or manipulation, and provide accurate information that empowers individuals to make informed decisions.

Case Study: Balancing Individual Rights and Public Health in Vaccine Communication

A vivid example of balancing individual rights and public health in communication arises in the context of vaccine promotion. Vaccination is a highly effective public health intervention, but it also raises questions about individual autonomy and informed consent.

Public health campaigns aiming to promote vaccination must strike a delicate balance. They need to respect individuals' right to make informed choices about their health while emphasizing the importance of vaccination for both personal and public health. Public health authorities must use evidence-based messaging that presents accurate information, addresses concerns, and highlights the potential benefits of vaccination.

Moreover, respecting individual privacy in vaccine communication necessitates securing personal information appropriately. Public health agencies must collect only the necessary information for vaccination records, ensure the confidentiality and secure storage of data, and transparently communicate data handling practices to maintain public trust.

Conclusion

Balancing individual rights to privacy, freedom of information, and autonomy with the need to protect public health is a complex ethical challenge, particularly during public health emergencies. Communicating effectively while respecting individual rights requires transparency, tailored messaging, consent options, and a commitment to accurate and accessible information.

By acknowledging the importance of individual rights and public health protection, public health agencies can navigate these ethical dilemmas and foster trust and collaboration with the public. Striking the delicate balance between individual rights and public health is crucial for promoting a collective response to public health emergencies and achieving a healthy and equitable future for all.

Ethical Considerations in Crisis Communication

Crisis communication plays a crucial role in managing public health emergencies and other crisis situations. It involves the dissemination of accurate and timely information to the public, decision-makers, and other stakeholders. However, ethical considerations must guide crisis communication practices to ensure transparency, fairness, and the protection of individual rights. In this section, we will explore some of the ethical issues that arise in crisis communication and discuss strategies for addressing them.

Transparency and Accuracy

One of the primary ethical considerations in crisis communication is the need for transparency and accuracy. It is essential to provide the public with truthful and up-to-date information about the crisis, including its causes, effects, and recommended actions. Transparency builds trust and credibility, and helps individuals make informed decisions regarding their health and safety.

However, achieving transparency can be challenging, especially when dealing with rapidly evolving situations and uncertainties. It is crucial to strike a balance between providing accurate information and avoiding premature or speculative statements that may cause panic or misinformation. Crisis communication teams must base their messages on reliable data and scientific evidence, clearly distinguishing between what is known, what is uncertain, and what is being done to address the crisis.

Balancing Individual Rights and Public Health

Another ethical consideration in crisis communication is the delicate balance between protecting individual rights and promoting public health. During a crisis, it may be necessary to implement measures such as quarantines, travel restrictions, or mandatory vaccinations to protect the population's health. Communicating these measures effectively while respecting individual autonomy and privacy is crucial.

Crisis communication should address the rationale behind these measures, explaining how they contribute to public health goals and minimize harm. It is essential to provide clear information on the rights and responsibilities of individuals during a crisis, including any potential limitations on their freedoms. Respecting privacy and confidentiality in collecting and sharing personal health information is also critical, ensuring that individuals' trust is maintained.

Addressing Public Anxiety and Emotional Distress

Crisis situations often evoke fear, anxiety, and emotional distress among the public. Ethical crisis communication should take into account these emotional reactions and address them with empathy and sensitivity. Providing psychological support, resources for coping, and avenues for seeking help are essential components of ethical crisis communication.

Additionally, crisis communication should anticipate and address potential rumors, misinformation, and stigmatization associated with the crisis. Promoting accurate information through multiple channels, engaging with credible sources, and debunking myths can help mitigate fear and minimize the harmful impact of misinformation on individuals' well-being.

Community Engagement and Participation

Ethical crisis communication should also value community engagement and participation. Engaging with the community affected by the crisis, as well as relevant stakeholders, enhances the effectiveness and fairness of communication efforts. Listening to community concerns, incorporating local knowledge, and involving affected individuals in decision-making processes foster trust and enable the development of contextually relevant communication strategies.

In crisis communication, cultural and linguistic considerations are crucial to ensure that information reaches diverse communities and is understood effectively. Recognizing and addressing disparities in access to information and resources is essential for ethical crisis communication.

Evaluation and Learning

Lastly, ethical crisis communication should include a commitment to ongoing evaluation and learning. The effectiveness of communication strategies should be assessed to identify areas for improvement and to ensure that messages are reaching the target audience. Feedback from the community and stakeholders should be sought and incorporated into future communication efforts.

Ethical crisis communication does not end when the crisis abates. It involves an ongoing process of learning from experiences, adapting to new challenges, and continuously improving communication practices for future crises.

Conclusion

Ethical considerations are at the core of crisis communication during public health emergencies and other crisis situations. Transparency, accuracy, balancing individual rights and public health, addressing public anxiety and emotional distress, community engagement and participation, and evaluation and learning are key principles that guide ethical crisis communication. By upholding these principles, crisis communication can foster trust, promote informed decision-making, and contribute to effective crisis management.

Ethical Implications of Mandatory Vaccination

Individual Autonomy and Vaccine Mandates

In the context of public health, individual autonomy refers to the right of individuals to make decisions about their own bodies and health. This principle is based on the respect for personal freedom and the belief that individuals should have control over what happens to their bodies. However, when it comes to vaccine mandates, the tension between individual autonomy and public health interests arises.

Vaccine mandates are laws or policies that require individuals to receive certain vaccines or prove their immunization status in order to access certain services, such as attending schools, participating in certain occupations, or traveling to certain countries. The goal of vaccine mandates is to protect public health by ensuring high vaccination rates and thereby preventing the spread of vaccine-preventable diseases.

Proponents of vaccine mandates argue that they are necessary to protect the vulnerable members of society, such as those who cannot receive vaccines due to medical reasons or those with weakened immune systems. They also argue that the benefits of vaccines outweigh the risks, and that by ensuring high vaccination rates, vaccine mandates contribute to the overall health and well-being of communities.

However, opponents of vaccine mandates raise concerns about individual autonomy and the infringement of personal rights. They argue that individuals should have the right to make decisions about their own bodies, including whether or not to receive vaccines. They also express concerns about the potential risks and side effects of vaccines, and question the evidence supporting their safety and efficacy.

Finding a balance between individual autonomy and the public health benefits of vaccines is a complex ethical issue. It requires careful consideration of the principles of autonomy, beneficence, non-maleficence, and justice.

One approach to addressing this tension is to prioritize individual autonomy while also promoting vaccine education and awareness. By providing individuals with accurate and reliable information about vaccines, their risks and benefits, and the potential consequences of non-vaccination, individuals can make informed decisions about their own health. This approach respects individual autonomy while also promoting public health.

Another approach is to implement vaccine mandates with certain exemptions and accommodations. This allows for individuals who have legitimate medical reasons for not receiving vaccines to be exempted, while still ensuring high vaccination rates in the general population. However, the criteria for exemptions should be carefully defined and based on scientific evidence to prevent abuse and maintain public health goals.

Moreover, it is important to address the underlying factors that contribute to vaccine hesitancy and mistrust. This could involve improving vaccine communication strategies, addressing misinformation, and building trust between healthcare providers and individuals. By addressing the root causes of vaccine hesitancy, it may be possible to increase vaccination rates without necessarily resorting to mandates.

In conclusion, the tension between individual autonomy and vaccine mandates is a complex ethical issue. While individual autonomy is a fundamental principle, it must be balanced with the public health benefits of vaccines. Finding a balance requires considering the principles of autonomy, beneficence, non-maleficence, and justice. Approaches that prioritize informed consent, provide exemptions based on legitimate medical reasons, and address underlying factors contributing to vaccine hesitancy can help navigate this ethical dilemma. It is important to continue the dialogue surrounding vaccine mandates and explore ways to promote vaccine acceptance while respecting individual autonomy.

Public Health Benefits and Risks of Mandatory Vaccination

Vaccination is one of the most effective public health interventions that has saved millions of lives and prevented the spread of infectious diseases. Mandatory vaccination policies have been implemented in many countries to ensure high vaccination rates and protect public health. However, these policies have also raised ethical questions regarding individual autonomy, freedom of choice, and potential risks associated with vaccines. In this section, we will explore the public health benefits of mandatory vaccination and discuss the potential risks and ethical considerations associated with it.

Public Health Benefits of Mandatory Vaccination

1. **Disease Prevention:** Mandatory vaccination plays a crucial role in preventing the spread of infectious diseases. By ensuring high vaccination rates in the population, herd immunity can be achieved, which provides indirect protection to those who cannot receive the vaccine due to medical conditions or age. This not only prevents the transmission of diseases but also reduces the overall disease burden in the community.

2. **Control of Outbreaks:** Mandatory vaccination policies are particularly important in controlling outbreaks of vaccine-preventable diseases. During outbreaks, communities with low vaccination rates are more susceptible to the rapid spread of infections. By mandating vaccination, these policies help contain and control outbreaks, protecting vulnerable populations and minimizing the impact on public health infrastructure.

3. **Protection of Vulnerable Populations:** Vaccination not only protects individuals who receive the vaccine but also helps to safeguard vulnerable populations such as the elderly, infants, pregnant women, and individuals with weakened immune systems. By ensuring high vaccination rates through mandatory policies, these populations are less likely to be exposed to vaccine-preventable diseases, reducing their risk of severe illness or complications.

4. **Long-Term Eradication:** Mandatory vaccination has played a crucial role in the eradication of diseases such as smallpox and the near-eradication of polio. These achievements demonstrate the long-term benefits of mandatory vaccination

policies, which contribute to global health security and eliminate the burden of certain diseases for future generations.

5. **Cost-Effectiveness:** Vaccination is a cost-effective public health intervention. By preventing the occurrence and spread of diseases, mandatory vaccination reduces the burden on healthcare resources, including hospitalizations, outpatient visits, and treatment costs. This results in long-term economic benefits for both individuals and society.

Risks and Ethical Considerations

1. **Individual Autonomy and Freedom of Choice:** One of the key ethical considerations regarding mandatory vaccination is the infringement on individual autonomy and the freedom to make medical decisions. Critics argue that individuals should have the right to refuse vaccination based on personal beliefs, religious reasons, or concerns about potential side effects. Balancing individual autonomy with public health benefits is a complex ethical challenge.

2. **Trust and Vaccine Hesitancy:** Mandatory vaccination policies can lead to a decline in public trust in vaccines and the healthcare system. Concerns about vaccine safety, perceived conflicts of interest, and misinformation can contribute to vaccine hesitancy. It is essential to address these concerns through transparent communication, education campaigns, and ensuring the safety and efficacy of vaccines.

3. **Risks of Adverse Events:** Although vaccine side effects are generally rare and outweighed by the benefits of vaccination, there is a small risk of adverse events. Mandatory vaccination policies necessitate a careful balance between protecting public health and recognizing potential risks. Continual monitoring of vaccine safety and transparent reporting of adverse events are crucial in addressing these concerns.

4. **Equity and Access:** Mandatory vaccination policies should consider equity and access to vaccines. Socioeconomic factors, geographic disparities, and marginalized communities may face barriers to vaccination, such as lack of healthcare access or vaccine affordability. Efforts should be made to ensure that these policies do not disproportionately burden disadvantaged populations and that access to vaccination is equitable.

5. **Exceptions and Exemptions:** Mandatory vaccination policies should allow for reasonable exemptions for medical reasons, such as allergies or immunocompromised individuals who cannot receive certain vaccines. However, exemptions based on personal belief or philosophical objections raise ethical questions and challenges in maintaining public health goals.

6. **Legal and Policy Considerations:** Implementing mandatory vaccination policies requires careful legal and policy considerations, taking into account national and international human rights frameworks. Striking a balance between protecting public health and respecting individual rights is essential in designing effective and ethical policies.

Case Study: Measles Outbreak and Mandatory Vaccination

In recent years, several countries have experienced significant measles outbreaks, highlighting the importance of high vaccination rates and the challenges posed by vaccine hesitancy. One example is the Disneyland measles outbreak in 2014, which started in a theme park and quickly spread due to low vaccination rates in the community.

To control the outbreak and prevent further spread, California implemented a mandatory vaccination law, removing non-medical exemptions for school children. The policy aimed to protect public health, particularly vulnerable populations who could not receive the vaccine. While the law faced legal challenges and ethical debates, it led to an increase in vaccination rates and helped contain the outbreak.

This case study demonstrates the necessity of mandatory vaccination policies in controlling outbreaks and protecting public health. It also highlights the ethical considerations and challenges involved in balancing individual autonomy, public health benefits, and equitable access to vaccines.

Conclusion

Mandatory vaccination policies are a valuable tool in protecting public health, preventing the spread of infectious diseases, and ensuring the well-being of vulnerable populations. While these policies raise ethical questions regarding individual autonomy and potential risks, the public health benefits are well-documented. Striking a balance between individual rights and public health goals is crucial in designing effective and ethical mandatory vaccination policies. Transparent communication, addressing vaccine hesitancy, equitable access, and continued monitoring of vaccine safety are key components in creating a sustainable and just health and wellness future.

Addressing Vaccine Hesitancy and Building Vaccine Confidence

Vaccine hesitancy refers to the delay in acceptance or refusal of vaccines despite the availability of vaccination services. It has become a significant global health concern, as it poses a threat to the success of immunization programs and the control of vaccine-preventable diseases. Building vaccine confidence is crucial to increase vaccine uptake and ensure public health protection. In this section, we will explore the ethical and practical considerations in addressing vaccine hesitancy and strategies for building vaccine confidence.

Understanding Vaccine Hesitancy

Vaccine hesitancy is a complex phenomenon influenced by various factors, including individual beliefs, cultural perspectives, social influences, and distrust in the healthcare system. It is important to recognize that vaccine hesitancy does not necessarily stem from a lack of knowledge or education, but rather from the perceived risks and benefits associated with vaccines.

To effectively address vaccine hesitancy, a comprehensive understanding of the underlying concerns is crucial. This requires engaging with vaccine-hesitant individuals and communities, listening to their anxieties, and addressing their specific questions and fears related to vaccines.

Addressing Vaccine Hesitancy Ethically

When addressing vaccine hesitancy, it is essential to uphold ethical principles such as respect for autonomy, beneficence, and justice. The following strategies can help navigate ethical considerations in the context of vaccine hesitancy:

1. **Providing accurate information:** Transparency and open communication are essential to address vaccine hesitancy. Healthcare professionals should provide evidence-based information about vaccine safety, efficacy, and the importance of vaccination. This includes addressing common misconceptions and myths associated with vaccines.

2. **Respecting autonomy and informed decision-making:** Individuals have the right to make decisions about their health, including whether to receive vaccines. Respecting their autonomy requires providing comprehensive and unbiased information and supporting informed decision-making. Healthcare professionals should engage in shared decision-making processes, ensuring that individuals' values and concerns are taken into account.

3. **Addressing inequalities in access:** Vaccine hesitancy may be more prevalent in marginalized communities due to historical injustices, lack of access to healthcare, or mistrust. To address vaccine hesitancy ethically, it is crucial to

identify and mitigate barriers to vaccine access and ensure equitable distribution.

Strategies for Building Vaccine Confidence

Building vaccine confidence is a multifaceted process that involves a combination of strategies targeting individuals, communities, and healthcare systems. The following strategies can help promote vaccine confidence:

1. **Enhancing vaccine literacy:** Improving vaccine literacy empowers individuals to make informed decisions and combat misinformation. Public health campaigns should focus on providing accessible and accurate information about vaccines, their benefits, and the potential risks associated with vaccine-preventable diseases.

2. **Strengthening trust in healthcare providers:** Healthcare providers play a crucial role in building vaccine confidence. By fostering effective communication, actively listening to concerns, and addressing them transparently, providers can establish trust and credibility with patients. It is also essential to acknowledge and empathize with individuals' fears and anxieties, taking their experiences and perspectives into account.

3. **Engaging communities:** Engaging with communities can mitigate vaccine hesitancy. Community leaders and influencers can play a crucial role in advocating for vaccines. Collaborating with community organizations, religious leaders, and trusted voices can help create a supportive environment for vaccination.

4. **Leveraging social media and digital platforms:** Social media and digital platforms have become prominent sources of health information. It is vital to leverage these platforms effectively to provide accurate information, counter misinformation, and engage with vaccine-hesitant individuals.

5. **Promoting vaccine mandates with ethical considerations:** While respecting autonomy is crucial, there are situations where vaccine mandates may be necessary to protect public health. Implementing vaccine mandates should be done carefully, considering ethical principles and ensuring equitable access to vaccines. Additionally, clear exemptions should be provided for individuals who have valid medical, religious, or philosophical objections.

Real-World Example: The COVID-19 Vaccine

The COVID-19 pandemic has highlighted the importance of addressing vaccine hesitancy. To ensure the successful rollout of COVID-19 vaccines, health authorities and policymakers must implement strategies to build vaccine confidence. This includes transparent communication about vaccine development, safety, and efficacy, addressing concerns related to the speed of vaccine development, and countering misinformation circulating on social media platforms.

Furthermore, community engagement, particularly with vulnerable populations and those who have historically experienced healthcare disparities, is crucial. By partnering with community organizations, healthcare providers can build trust, address cultural concerns, and ensure equitable access to vaccines.

Addressing vaccine hesitancy is an ongoing process that requires continuous monitoring, evaluation, and adaptation of strategies. By upholding ethical principles and implementing evidence-based interventions, healthcare systems can effectively address vaccine hesitancy, strengthen vaccine confidence, and promote public health.

Ethical Considerations in Health and Wellness Technology

Telemedicine and Virtual Healthcare

Ethical Issues in Remote Healthcare Delivery

In recent years, there has been a rapid advancement in technology that has transformed the field of healthcare. One of the most notable developments is the emergence of remote healthcare delivery, also known as telemedicine or telehealth. Remote healthcare delivery allows healthcare providers to diagnose, monitor, and treat patients from a distance using telecommunications technology. While this advancement has the potential to improve access to healthcare and enhance patient outcomes, it also raises several ethical concerns that need to be carefully considered.

Privacy and Security

One of the primary ethical issues in remote healthcare delivery is the protection of patient privacy and the security of their health information. When patients engage in telemedicine consultations, their personal health data is transmitted over electronic networks. This data may include sensitive information such as medical records, test results, and images. Therefore, it is crucial to ensure that appropriate safeguards are in place to protect the confidentiality and integrity of this information.

Healthcare providers must implement secure communication channels and employ encryption protocols to prevent unauthorized access or interception of patient data. Additionally, healthcare organizations should have robust security measures to safeguard against data breaches and cyber attacks. These measures

339

include regular system audits, updates, and employee training on best practices for data protection. It is also important to inform patients about the potential risks and benefits of remote healthcare delivery and obtain their informed consent prior to engaging in telemedicine services.

Quality of Care and Diagnostic Accuracy

Another ethical concern in remote healthcare delivery is maintaining the quality of care and ensuring accurate diagnoses. In traditional face-to-face healthcare encounters, healthcare providers can rely on direct physical examination and visual observation to assess a patient's condition. However, in telemedicine, healthcare providers must rely on the information provided by the patient, as well as any available diagnostic tests and imaging studies.

This reliance on patient-reported symptoms and remote examination techniques might introduce challenges in accurately diagnosing and treating certain conditions. It is crucial for healthcare providers to establish clear communication channels with patients to ensure accurate reporting of symptoms and complaints. The use of high-quality video conferencing and peripheral devices, such as digital stethoscopes and otoscopes, can enhance the diagnostic accuracy in remote consultations.

Furthermore, healthcare providers must be aware of the limitations and potential biases associated with remote healthcare delivery. Awareness of these limitations can help providers make informed decisions about when it is appropriate to use remote care versus in-person care. In situations where a physical examination is necessary, patients should be referred for in-person evaluations to ensure appropriate care and diagnosis.

Equitable Access and Technological Divide

Remote healthcare delivery has the potential to improve access to healthcare, particularly for individuals who face geographic, financial, or mobility barriers. However, it is essential to consider the potential disparities created by the technological divide.

Not all individuals have access to reliable internet connections, smartphones, or computers, which are necessary for engaging in telemedicine services. This lack of access can disproportionately affect individuals from low-income backgrounds or rural areas, exacerbating existing healthcare disparities. Additionally, older adults or individuals with limited digital literacy may struggle to navigate telemedicine platforms.

To address these inequities, healthcare organizations and policymakers should strive to ensure equitable access to remote healthcare services. This includes increasing access to internet infrastructure in underserved areas, providing technological devices to individuals who cannot afford them, and offering training programs to enhance digital literacy among patient populations.

Continuity of Care and Patient-Provider Relationship

Another ethical consideration in remote healthcare delivery is the preservation of the patient-provider relationship and continuity of care. In traditional healthcare settings, patients often have established relationships with their healthcare providers, which can foster trust, empathy, and effective communication. The introduction of remote interactions can potentially disrupt this relationship.

Healthcare providers must make efforts to build rapport and trust with their remote patients. This can be achieved through clear communication, active listening, and empathy. Providers should also ensure that there are mechanisms in place for follow-up visits and continuous monitoring to maintain the continuity of care.

Moreover, patients should have the autonomy to choose between remote care and in-person care. Shared decision-making should be encouraged, allowing patients to weigh the benefits and drawbacks of each modality and make informed choices based on their preferences and unique circumstances.

Resource Allocation and Health Inequities

Remote healthcare delivery can contribute to the optimization of healthcare resources by reducing the burden on in-person healthcare facilities. However, it is crucial to consider the potential impact on resource allocation and health inequities.

In some cases, remote care may lead to increased demand for healthcare services, as patients may be more likely to seek care due to the convenience and accessibility of telemedicine. This increased demand can strain healthcare resources, particularly in underserved communities or regions with limited healthcare infrastructure.

Healthcare organizations and policymakers should carefully evaluate the allocation of resources and ensure that the benefits of remote healthcare delivery are equitably distributed. This involves addressing the systemic inequities in healthcare access and infrastructure. It is also important to monitor and evaluate the impact of remote healthcare delivery on health outcomes to identify and address any unintended consequences or disparities.

Case Study: Remote Rehabilitation Services

To illustrate the ethical issues in remote healthcare delivery, let's consider a case study involving remote rehabilitation services. John, a 40-year-old construction worker, recently suffered a severe lower back injury and requires ongoing physical therapy for his recovery. Due to his limited mobility and the lack of nearby rehabilitation clinics, John's healthcare provider recommends remote rehabilitation services.

While remote rehabilitation services would enhance John's access to care, there are ethical considerations that need to be addressed. First, ensuring the privacy and security of John's personal health information is essential. John's provider should use secure communication channels and discuss the potential risks and benefits of remote rehabilitation services with him.

Second, maintaining the quality of care is crucial for effective rehabilitation. John's provider could request regular video consultations and use portable devices to monitor his progress. However, it is important to acknowledge the limitations of remote care and refer John for in-person evaluations when necessary.

Lastly, John's socioeconomic status may impact his ability to access remote rehabilitation services. The provider should assess John's technological capabilities and ensure that he has the necessary equipment and internet access. If John lacks the resources, the provider should explore alternative solutions or advocate for increased accessibility in his community.

By addressing these ethical issues, John's remote rehabilitation services can be delivered ethically and ensure that he receives comprehensive care and support during his recovery.

Conclusion

Remote healthcare delivery has the potential to revolutionize the healthcare landscape, improving access to services and enhancing patient outcomes. However, it is crucial to navigate the ethical concerns associated with this advancement. Ensuring patient privacy and security, maintaining the quality of care, addressing disparities in access, promoting continuity of care and the patient-provider relationship, and evaluating resource allocation are essential for the ethical implementation and practice of remote healthcare delivery.

As technology continues to advance, it is imperative that healthcare providers, policymakers, and society as a whole engage in ongoing dialogue and collaboration to establish ethical frameworks and guidelines that prioritize patient well-being and uphold the principles of justice and equity in remote healthcare delivery.

Equity and Access in Telemedicine

Telemedicine has emerged as a powerful tool in healthcare delivery, providing remote access to medical services and eliminating geographical barriers. However, as with any technological advancement, there are ethical considerations to be addressed, particularly in terms of equity and access. In this section, we will explore the ethical issues surrounding equity and access in telemedicine and discuss strategies for promoting fairness and inclusiveness in its implementation.

Understanding the Equity Gap in Telemedicine

Telemedicine has the potential to revolutionize healthcare by providing convenient and efficient access to medical care. However, it is important to acknowledge that not everyone has equal access to telemedicine services. This equity gap can arise due to various barriers, including technological, geographical, socioeconomic, and cultural factors.

Technological barriers may arise from limited access to reliable internet connections or a lack of necessary devices or equipment. Geographical barriers can be a significant challenge in rural or remote areas where infrastructure may not support telemedicine services. Socioeconomic factors, such as income disparities, can also contribute to unequal access to technology and resources. Additionally, cultural factors, including language barriers or lack of familiarity with digital platforms, may further exacerbate the equity gap in telemedicine.

Ethical Considerations in Promoting Equity and Access

Addressing the equity gap in telemedicine requires a multifaceted approach that takes into consideration ethical considerations. Here are some key ethical considerations in promoting equity and access in telemedicine:

1. **Fair allocation of resources:** It is essential to ensure that the distribution of telemedicine resources is equitable, taking into account the needs of underserved populations. This requires proactive efforts to identify and mitigate disparities in access to technology and infrastructure.

2. **Cultural competency:** To promote equitable access, telemedicine providers should strive to understand and address cultural and linguistic barriers that may exist. This includes providing language translation services, culturally sensitive care, and education to promote health literacy.

3. **Privacy and data security:** Maintaining patient privacy and data security are crucial ethical considerations in telemedicine. Clear policies and safeguards must be implemented to protect patient information and ensure confidentiality.

Additionally, patients should have control over their data and understand how it is being used.

4. **Informed consent:** Telemedicine encounters should adhere to the principles of informed consent. Patients must be adequately informed about the limitations, risks, and benefits of telemedicine services to make autonomous decisions about their healthcare.

5. **Mitigating biases:** Telemedicine platforms should be designed and implemented in a way that minimizes biases and discrimination. This includes addressing algorithmic biases and ensuring equal access to services irrespective of race, gender, ethnicity, or socioeconomic status.

Strategies for Promoting Equity and Access

Promoting equity and access in telemedicine requires dedicated efforts from various stakeholders, including policymakers, healthcare organizations, and technology developers. Here are some strategies that can be employed:

1. **Infrastructure development:** Investment in telecommunication infrastructure is essential, particularly in underserved areas, to ensure reliable internet connectivity and access to telemedicine services. Collaborations between public and private sectors can play a crucial role in expanding infrastructure.

2. **Patient education and empowerment:** Providing education and resources to patients to navigate telemedicine platforms can enhance access and promote health literacy. This includes addressing language barriers, providing clear instructions, and offering support for patients with limited technological proficiency.

3. **Training for healthcare professionals:** Healthcare professionals should receive training in telemedicine practices, including cultural competency and digital literacy, to ensure equitable and quality care. This can help address potential biases and enhance the patient-provider relationship.

4. **Collaborative partnerships:** Collaboration between healthcare organizations, technology companies, and community organizations can help bridge the equity gap in telemedicine. Such partnerships can provide resources, support, and expertise to address barriers and promote access.

5. **Policy and regulation:** Governments and regulatory bodies must develop policies and regulations that promote equitable access to telemedicine services. This may include incentives for healthcare organizations to serve underserved populations and regulations to ensure data privacy and security.

Case Study: Telemedicine in Rural India

A case study of telemedicine implementation in rural India highlights the importance of addressing equity and access. In regions with limited access to healthcare facilities, telemedicine has been instrumental in providing medical consultations remotely. However, challenges with technological infrastructure and technological literacy have hindered its widespread adoption.

To address these challenges, a non-profit organization partnered with the local government to establish telemedicine centers in rural areas. The initiative focused on training healthcare workers, providing necessary equipment, and creating awareness among the community. By investing in infrastructure and capacity building, the initiative successfully improved access to healthcare services, particularly for vulnerable populations.

Conclusion

Equity and access are key ethical considerations in the implementation of telemedicine. While telemedicine holds great promise for improving healthcare accessibility, it is vital to address the equity gap and ensure that marginalized and underserved populations can benefit from these services. By considering the ethical principles discussed and employing strategies aimed at promoting fairness and inclusiveness, we can create a more equitable and just future for telemedicine.

Privacy and Security Concerns in Telemedicine

As telemedicine becomes more prevalent in healthcare, it brings with it numerous advantages in terms of accessibility and convenience. However, along with these benefits come important privacy and security concerns that need to be addressed. In this section, we will explore the ethical considerations surrounding privacy and security in telemedicine and discuss strategies for ensuring the protection of patient information.

Importance of Privacy and Security in Telemedicine

Privacy and security are fundamental principles in healthcare, and they are equally crucial in telemedicine. Telemedicine involves the exchange of sensitive patient information over electronic platforms, such as video calls, messaging apps, or online portals. This information includes personal and medical data, which can be highly confidential and valuable to malicious actors if accessed without permission.

Maintaining privacy and security in telemedicine is essential for several reasons. First, it upholds patient confidentiality, which is a foundational principle in healthcare ethics. Patients need assurance that their personal information will be safeguarded to maintain trust in the healthcare system. Second, securing telemedicine platforms protects against unauthorized access and potential data breaches, preventing harm to patients and minimizing the risk of identity theft or fraud. Finally, privacy and security measures can help ensure compliance with legal and regulatory requirements related to patient data protection.

Challenges in Privacy and Security

While the benefits of telemedicine are evident, there are challenges and risks associated with privacy and security. These challenges include:

1. **Data encryption and transmission:** To protect patient data, telemedicine platforms must use robust encryption techniques to secure data transmission over the internet. However, ensuring consistent and strong encryption practices across various platforms and devices can be a challenge.

2. **Authentication and access control:** Verifying the identity of both healthcare providers and patients is crucial to prevent unauthorized access to telemedicine platforms. Implementing multi-factor authentication and strong access control mechanisms helps ensure that only authorized individuals can access patient information.

3. **Data storage and retention:** Telemedicine platforms must have secure storage systems to store patient data. Adequate measures must be in place to protect stored data from unauthorized access and to ensure compliance with data retention policies that vary across jurisdictions.

4. **Third-party involvement:** Telemedicine platforms often rely on third-party vendors for various services, such as data hosting or software development. It is essential to carefully evaluate these vendors' privacy and security practices and ensure that they adhere to ethical standards and regulatory requirements.

5. **Patient education and awareness:** Patients must be knowledgeable about the privacy and security risks associated with telemedicine. They should be informed about the measures in place to protect their data and understand their role in maintaining privacy and security, such as using secure connections and keeping login credentials confidential.

Strategies for Ensuring Privacy and Security

To address these challenges and ensure privacy and security in telemedicine, healthcare organizations and telemedicine platforms can implement the following strategies:

1. **Compliance with privacy and security regulations:** It is essential to understand and comply with applicable laws and regulations regarding patient data protection. This includes adhering to standards such as the Health Insurance Portability and Accountability Act (HIPAA) in the United States or the General Data Protection Regulation (GDPR) in the European Union.

2. **Robust encryption and data transmission protocols:** Telemedicine platforms should employ strong encryption algorithms and secure data transmission protocols to protect patient information during transmission. Regular audits and assessments can help ensure the effectiveness of these measures.

3. **Authentication and access controls:** Implementing multi-factor authentication and strong access controls helps prevent unauthorized access to patient data. This includes verifying the identity of healthcare providers and patients and granting access privileges based on the principle of least privilege.

4. **Secure data storage and retention:** Telemedicine platforms should utilize secure storage systems with access controls and regular data backups. Clear guidelines for data retention and proper disposal should be followed to protect patient privacy.

5. **Vendor assessment and oversight:** Healthcare organizations should perform due diligence when selecting and evaluating third-party vendors. Contracts should clearly define expectations for privacy and security practices, including regular security assessments and audits.

6. **Patient education and informed consent:** Patients should receive clear information about privacy and security practices related to telemedicine. Informed consent forms should detail the risks, benefits, and specific privacy and security measures in place. Patient education materials and resources can further enhance awareness and understanding.

Case Study: Addressing Privacy and Security Concerns in Telemedicine

Let's consider a case study where a telemedicine platform was found to have experienced a data breach, compromising the personal and medical information of thousands of patients. The breach occurred due to inadequate encryption and data storage practices, leading to unauthorized access to patient data by a malicious hacker.

To address this breach and prevent future incidents, the telemedicine platform took several actions. They immediately strengthened their encryption protocols and implemented end-to-end encryption for all communication channels to ensure secure data transmission. Additionally, they conducted a comprehensive security audit and implemented more robust access controls, including multi-factor authentication for healthcare providers and patients. The platform also improved their data storage systems, employing encrypted databases and regular backups to protect stored information.

Moreover, the platform enhanced their vendor oversight procedures, conducting thorough assessments of third-party vendors' privacy and security practices. They also implemented a mandatory privacy and security training program for all employees to raise awareness and ensure compliance with best practices.

Through these efforts, the telemedicine platform demonstrated their commitment to addressing privacy and security concerns in telemedicine, rebuilding patient trust, and safeguarding patient information.

Conclusion

Privacy and security concerns are of utmost importance in telemedicine. Protecting patient confidentiality and ensuring the security of sensitive data are essential to maintain trust in healthcare systems. By implementing robust encryption practices, strong authentication measures, and secure data storage systems, healthcare organizations can mitigate privacy and security risks in telemedicine. It is equally important to educate and empower patients, providing them with the knowledge and tools necessary to actively participate in maintaining privacy and security during telemedicine encounters. Through these ethical practices, the future of telemedicine can thrive while upholding patient privacy and security.

Artificial Intelligence and Machine Learning in Healthcare

Ethical Challenges in AI-driven Diagnostics and Treatment

The use of artificial intelligence (AI) in healthcare has the potential to revolutionize diagnostics and treatment, improving patient outcomes and enhancing the efficiency of medical care. AI algorithms can analyze large amounts of medical data, identify patterns, and make accurate predictions, enabling faster and more accurate diagnoses. AI-driven technologies can also assist healthcare providers in treatment planning, enabling personalized and targeted therapies. However, the integration of AI in healthcare also presents several ethical challenges that need to be carefully considered and addressed. In this section, we will explore the ethical challenges associated with AI-driven diagnostics and treatment and discuss possible solutions.

Privacy and data security

One of the biggest ethical concerns inherent in AI-driven diagnostics and treatment is the privacy and security of patient data. AI algorithms rely on vast amounts of personal health information to train and make accurate predictions. While this data is crucial for the development of AI models, it also poses significant risks to patient privacy if not handled properly.

To protect patient privacy, healthcare organizations must ensure that robust security measures are in place to prevent unauthorized access, breaches, or misuse of patient data. Encryption techniques, access controls, and secure storage systems can be employed to safeguard the confidentiality and integrity of patient information. Additionally, stringent data anonymization techniques should be used to minimize the risk of re-identification.

Ethical considerations also extend to the sharing of patient data with third-party vendors or researchers. Transparent and informed consent processes should be implemented, giving patients control over how their data is used and shared. It is essential to establish trust between patients, healthcare providers, and AI developers to ensure that patient data is used solely for the intended purposes and with proper safeguards.

Bias and fairness

AI algorithms are only as good as the data they are trained on. If training datasets are biased or unrepresentative, AI models can perpetuate and amplify existing biases, leading to unfair treatment and healthcare disparities. This presents a significant ethical challenge in AI-driven diagnostics and treatment.

To address bias in AI algorithms, it is crucial to ensure diverse and representative training datasets. This includes not only demographic diversity but also diversity across different geographical locations, socioeconomic backgrounds, and disease prevalence. Care must be taken to prevent underrepresentation or marginalization of specific populations, which could result in unequal access to accurate diagnoses and treatments.

Regular evaluation and monitoring of AI algorithms for bias and fairness are essential. Bias detection methods, such as statistical analysis and fairness metrics, should be employed to identify and mitigate biases in real-time. Transparency in the development and deployment of AI models can help improve accountability and ensure that biases are not inadvertently perpetuated.

Lack of human oversight and accountability

While AI algorithms can provide valuable insights and recommendations, they should not replace human judgment and oversight. The lack of human involvement in the decision-making process raises concerns about accountability and responsibility in AI-driven diagnostics and treatment.

Healthcare providers must retain ultimate responsibility for patient care and treatment decisions. AI algorithms should be viewed as decision support tools, assisting healthcare providers in interpreting medical data and making informed decisions. It is crucial to establish clear guidelines on the roles and responsibilities of AI systems and humans to ensure that AI technologies are used ethically and appropriately.

Additionally, the transparency of AI algorithms is vital for accountability. Healthcare providers and patients should have access to information about how AI models work, what data they rely on, and how their predictions or recommendations are generated. This transparency promotes trust in AI systems and enables critical evaluation of their outputs.

Explainability and interpretability

AI algorithms often operate as black boxes, making it challenging to understand how they arrive at specific diagnoses or treatment recommendations. The lack of explainability and interpretability can be an ethical concern in healthcare, as it makes it difficult for healthcare providers and patients to trust and validate the outputs of AI algorithms.

To address this challenge, efforts should be made to develop AI models that are explainable and interpretable. Explainable AI techniques, such as model-agnostic interpretability methods and rule-based systems, can provide insights into the decision-making process of AI algorithms. These techniques enable healthcare providers to understand and validate the reasoning behind the recommendations made by AI systems.

It is essential to strike a balance between model complexity and interpretability. While complex AI models can achieve high accuracy, they may lack interpretability. Simpler models that sacrifice some accuracy for interpretability may be more ethically appropriate, as they allow for better understanding and evaluation of the decision-making process.

Informed consent and patient autonomy

Informed consent is a fundamental ethical principle in healthcare, ensuring that patients have the right to make autonomous decisions about their medical care. The integration of AI in diagnostics and treatment raises questions about how informed consent can be obtained and maintained.

Healthcare providers must ensure that patients understand the implications of AI-driven diagnostics and treatment, including the limitations and risks associated with these technologies. Patients should have the opportunity to ask questions, seek clarifications, and provide their consent based on a comprehensive understanding.

While AI algorithms can provide valuable insights, the final decisions about treatment should always involve shared decision-making between healthcare providers and patients. AI should be used as a tool to support patient autonomy, ensuring that patients are actively involved in their healthcare journey.

Mitigating unintended consequences

The deployment of AI in healthcare may have unintended consequences that need to be carefully considered and mitigated. For example, the reliance on AI may lead to the deskilling of healthcare providers, reducing their ability to make independent judgments. Similarly, overreliance on AI may result in medical errors if the technology fails or produces inaccurate predictions.

Stringent quality control measures should be in place to detect and rectify any errors or glitches in AI algorithms. Continuous monitoring and evaluation of AI systems are essential to ensure their reliability, accuracy, and safety. Education and training programs for healthcare providers can equip them with the necessary skills to understand, use, and critically evaluate AI technologies.

Moreover, a commitment to ongoing research and development is crucial to advance the capabilities, safety, and ethical use of AI in healthcare. Collaboration between AI developers, healthcare providers, and ethicists is essential to create a framework that addresses the ethical challenges posed by AI-driven diagnostics and treatment.

Conclusion

AI-driven diagnostics and treatment offer great promise in improving healthcare outcomes. However, the ethical challenges discussed in this section must be addressed to ensure the responsible and equitable use of AI technologies. Privacy and data security, bias and fairness, lack of human oversight, explainability and interpretability, informed consent and patient autonomy, and the mitigation of unintended consequences are crucial considerations for ethically integrating AI into healthcare. By carefully addressing these challenges, we can harness the potential of AI to advance diagnostic accuracy, optimize treatment decisions, and ultimately improve patient care.

Bias and Discrimination in AI Algorithms

In recent years, artificial intelligence (AI) has made significant advancements in various domains, including healthcare, finance, and education. AI algorithms are designed to make decisions and predictions based on large amounts of data. However, these algorithms are not immune to biases and can inadvertently perpetuate and amplify existing biases and discrimination present in society. This section explores the ethical implications of bias and discrimination in AI algorithms and discusses strategies to mitigate these issues.

Understanding Bias in AI Algorithms

Bias in AI algorithms can arise due to a variety of factors, including biased training data, biased features, biased model selection, and biased interpretations. It is important to understand that biases in AI algorithms are not inherent to the technology itself but rather a reflection of the biases present in the data used to train and develop these algorithms.

Biased Training Data: AI algorithms learn from large datasets, and if the training data contains biases, the algorithm will learn and reproduce those biases in its decision-making process. For example, if the training data for a loan approval system predominantly consists of approved loans for a particular demographic group, the algorithm may disproportionately favor that group in its loan approval decisions.

Biased Features: The features used by AI algorithms to make predictions can also introduce bias. For instance, if an algorithm that predicts job performance relies heavily on a candidate's education level, it may disproportionately favor candidates from privileged backgrounds who have had better access to education.

Biased Model Selection: The process of selecting and fine-tuning AI models can introduce bias. The biases of the individuals involved in developing the AI system can influence the design and decision-making process, leading to biased outcomes.

Biased Interpretations: Bias can also arise in the interpretation of AI-generated results. Human interpreters may selectively focus on certain findings or patterns, leading to biased interpretations and decision-making.

Ethical Implications of Bias in AI Algorithms

The presence of bias in AI algorithms has significant ethical implications. Biased AI algorithms can perpetuate and amplify societal biases and discrimination, exacerbating existing inequalities. These biases can manifest in various ways, including but not limited to:

Discrimination: Biased AI algorithms can discriminate against individuals or groups based on various protected attributes, such as race, gender, age, or sexual orientation. For example, biased facial recognition technology may disproportionately misidentify people of certain races, leading to false accusations or biased surveillance.

Unfair Resource Allocation: Biased AI algorithms may lead to the unfair allocation of resources, such as loans, healthcare interventions, or educational opportunities. If the algorithm perpetuates systemic biases, it may deny resources or opportunities to marginalized individuals or groups, further entrenching inequalities.

Reinforcement of Stereotypes: Biased AI algorithms can reinforce and perpetuate stereotypes by associating certain attributes or behaviors with specific demographic groups. This can have harmful effects on individuals by limiting their opportunities based on inaccurate assumptions.

Lack of Accountability: Biased AI algorithms can make discriminatory decisions without accountability, as they are often viewed as objective and impartial. This lack of transparency and accountability can make it challenging to address and rectify biased outcomes.

Strategies to Mitigate Bias and Discrimination in AI Algorithms

Addressing bias and discrimination in AI algorithms requires a multifaceted approach involving data collection, algorithm design, and continued monitoring. Here are some strategies to mitigate bias and discrimination:

Diverse and Representative Training Data: Ensuring that training data is diverse and representative of the populations the algorithm will be used on can help mitigate bias. This requires careful consideration of the sources and collection methods of the data to avoid perpetuating existing biases.

Effective Feature Selection: Thoughtful selection of features in AI algorithms can help avoid incorporating biased attributes into decision-making processes. It is crucial to critically examine the relevance and fairness of each feature and assess the potential for bias.

Algorithmic Fairness: Researchers and developers should incorporate fairness considerations into the design of AI algorithms. This includes exploring fairness metrics, such as equal opportunity, equalized odds, and demographic parity, to measure and address bias in algorithmic decision-making.

Interpretability and Transparency: Enhancing the interpretability and transparency of AI algorithms can help identify and rectify biases. Techniques such as interpretable machine learning models and algorithmic audits can contribute to better understanding and accountability.

Ongoing Monitoring and Evaluation: Regular monitoring and evaluation of AI algorithms' outcomes and impacts are vital to identify and correct any biases that may arise over time. This includes collecting feedback from affected individuals or groups and implementing remedial actions when biases are identified.

Real-World Examples

Biased AI algorithms have received considerable public attention in recent years. Some real-world examples include:

Facial Recognition Technology: Facial recognition technology has been shown to exhibit bias, particularly in misidentifying individuals from certain racial or ethnic backgrounds. This bias can lead to harmful consequences, such as false arrests or biased surveillance targeting specific communities.

Predictive Policing: AI algorithms used in predictive policing have been criticized for perpetuating racial biases and leading to over-policing of certain communities. Biased algorithms can result in disproportionate surveillance and arrests, further exacerbating systemic inequalities in the criminal justice system.

Automated Hiring Systems: AI-powered hiring systems have been found to exhibit biases in favor of certain demographic groups. Biased algorithms can perpetuate gender, racial, or age discrimination, leading to unequal employment opportunities.

Conclusion

Addressing bias and discrimination in AI algorithms is essential to ensure that they contribute to a fair and just society. It requires a collaborative effort from researchers, policymakers, and technology developers. By acknowledging and understanding the ethical implications of bias in AI algorithms and implementing strategies to mitigate these biases, we can strive towards the development of more equitable and unbiased AI systems that promote justice and equality.

Accountability and Transparency in AI Systems

In recent years, artificial intelligence (AI) has become increasingly prevalent in various fields, including healthcare. AI systems can greatly assist in tasks such as diagnostics, treatment planning, and predictive modeling. However, the rapid advancement of AI has raised concerns about the accountability and transparency of these systems. In this section, we will explore the ethical and practical considerations related to accountability and transparency in AI systems in healthcare.

The Importance of Accountability

Accountability is essential in AI systems to ensure that the outcomes produced by these systems are reliable, unbiased, and ethical. In healthcare, where AI algorithms can impact patient care and decisions, accountability becomes even more crucial. Accountability helps to establish trust in AI systems and protects the rights and well-being of patients.

Transparency in AI Systems

One aspect of accountability is transparency. Transparent AI systems provide clear explanations of how they arrive at their decisions or recommendations. Openness and transparency allow healthcare professionals and patients to understand and evaluate the AI system's outputs, leading to better-informed decision-making.

There are several ways to achieve transparency in AI systems. One approach is to provide interpretability, where the AI system makes its decision-making process understandable to humans. This can be done by using interpretable machine learning models, providing visualizations of the internal workings of the AI system, or generating explanations for its outputs.

Another approach is to disclose the data used to train the AI system, as well as any biases present in the dataset. Transparency in data collection and use is crucial to ensure fairness and to address potential biases in the AI system's decision-making.

Challenges and Solutions

Achieving accountability and transparency in AI systems in healthcare comes with its own set of challenges. One challenge is the complexity of AI algorithms, which often operate as a black box, making it difficult to understand their inner workings. This lack of interpretability hinders the ability to identify and address biases or errors in the system.

To overcome this challenge, researchers have been working on developing explainable AI (XAI) techniques. XAI aims to create AI systems that provide clear explanations of their outputs, making them more transparent and understandable to both healthcare professionals and patients. These techniques include visual explanations, rule-based models, and counterfactual explanations, among others.

Another challenge is the availability of high-quality and diverse datasets. Biases in healthcare data can lead to biased AI systems that perpetuate inequalities or make incorrect decisions. To address this challenge, efforts should be made to collect comprehensive and diverse datasets that accurately represent different populations. Moreover, transparency in data collection practices is necessary to identify and mitigate biases.

Collaboration between AI developers, healthcare professionals, and ethics experts can also help tackle the challenges of accountability and transparency. By involving various stakeholders in the development and evaluation of AI systems, a more comprehensive understanding of the ethical implications and potential biases can be achieved.

Real-World Examples

To illustrate the importance of accountability and transparency in AI systems, let's consider a real-world example. Imagine an AI system designed to predict the risk of cardiovascular disease. If this system lacks transparency, healthcare professionals may not have a clear understanding of why certain patients are being classified as high-risk or low-risk. This lack of transparency could lead to hesitation in implementing interventions or inappropriate treatment decisions. However, if the AI system is transparent and provides explanations for its predictions, healthcare professionals can better understand the reasoning behind the risk classifications and make more informed decisions.

Resources and Further Reading

For those interested in delving deeper into the topic of accountability and transparency in AI systems, here are some recommended resources:

- Book: "Ethics of Artificial Intelligence and Robotics" by Vincent C. Müller.

- Journal Article: "Explainable AI: Why We Need It and How to Make It Happen" by Been Kim et al., in Communications of the ACM.

- Online Resource: Oxford Internet Institute's "Ethical Impact Assessment of Algorithms in Healthcare".

Conclusion

Accountability and transparency are crucial aspects of AI systems in healthcare. By ensuring that AI systems are accountable and transparent, we can enhance trust, fairness, and understanding in the use of AI for healthcare decision-making. Striving for accountability and transparency in AI systems will ultimately contribute to more ethical and just health and wellness practices.

Personal Health Monitoring and Genetic Testing

Privacy and Security of Personal Health Data

In the digital age, the collection and storage of personal health data have become integral to the delivery of healthcare services and the advancement of medical research. However, the increasing reliance on technology raises concerns about the privacy and security of personal health data. This section explores the ethical considerations surrounding the privacy and security of personal health data and provides insights into the challenges and solutions in this domain.

The Significance of Personal Health Data Privacy

Personal health data includes sensitive information such as medical history, test results, and genetic information. Protecting the privacy of this data is essential for several reasons:

+ **Confidentiality and Trust:** Patients have a legitimate expectation that their personal health information will be kept confidential. Respect for privacy fosters trust between healthcare providers and patients, ensuring open and honest communication.

+ **Autonomy and Control:** Individuals should have control over their personal health data, including who has access to it and how it is used. Maintaining privacy empowers individuals to make informed decisions about their healthcare.

+ **Non-Discrimination:** Personal health information can be used to discriminate against individuals based on their health status. Protecting privacy helps prevent unfair treatment based on health-related information.

+ **Research and Public Interest:** Privacy protections encourage individuals to participate in medical research and public health initiatives. This promotes the advancement of scientific knowledge and the overall well-being of society.

Challenges in Privacy and Security

Ensuring the privacy and security of personal health data is not without its challenges. The following are some of the key issues in this domain:

1. **Data Breaches:** The digitization of health records creates vulnerabilities that can be exploited by hackers. Data breaches can result in the unauthorized disclosure or use of personal health information, leading to identity theft, insurance fraud, and other forms of harm.

2. **Insider Threats:** Healthcare professionals and other authorized personnel may misuse or disclose personal health data for personal gain or other unethical purposes. Protecting against insider threats requires robust security measures and strong ethical guidelines.

3. **Data Aggregation and Secondary Use:** Personal health data is often aggregated and used for purposes beyond the original intent, such as

research or marketing. Ensuring transparency and obtaining explicit consent for secondary uses is essential to maintain privacy.

4. **Emerging Technologies:** Advancements in technologies like artificial intelligence and machine learning raise concerns about the potential misuse or unauthorized access to personal health data. Ethical considerations must be addressed to harness the benefits of these technologies without compromising privacy.

Ethical Principles and Frameworks

Several ethical principles and frameworks can guide the protection of personal health data:

+ **Privacy:** Individuals have the right to control the collection, use, and disclosure of their personal health data. Privacy principles provide a foundation for respecting this right and ensuring that data is only accessed by authorized entities.

+ **Security:** Robust security measures are necessary to protect personal health data from unauthorized access, data breaches, and other forms of harm. Security protocols should be regularly updated to adapt to evolving threats.

+ **Informed Consent:** Individuals should be adequately informed about the collection, use, and sharing of their personal health data. Informed consent ensures that individuals have the necessary information to make autonomous decisions about their data.

+ **Minimization and De-identification:** Collecting and storing only the necessary data minimizes privacy risks. De-identifying personal health data before it is used for secondary purposes can further protect privacy.

+ **Accountability:** Entities that collect and handle personal health data should be accountable for their actions. This includes regularly reviewing and auditing security measures, promptly responding to breaches, and ensuring compliance with privacy regulations.

Legal and Regulatory Frameworks

Various legal and regulatory frameworks aim to protect personal health data privacy and security. Examples include:

+ The **Health Insurance Portability and Accountability Act (HIPAA)** in the United States establishes standards for the protection and privacy of sensitive health data.

+ The **General Data Protection Regulation (GDPR)** in the European Union mandates the protection of personal data, including health information, and gives individuals greater control over their data.

+ The **Personal Information Protection and Electronic Documents Act (PIPEDA)** in Canada regulates the collection, use, and disclosure of personal health information.

Complying with these legal frameworks is crucial for healthcare organizations and other entities handling personal health data.

Strategies for Protecting Personal Health Data

To address the challenges involved in protecting personal health data, several strategies can be employed:

+ **Encryption and Access Controls:** Protecting personal health data through encryption techniques and access controls can prevent unauthorized access and mitigate the impact of data breaches.

+ **Anonymization and De-identification:** Removing identifying information from datasets can help protect individuals' privacy while allowing for meaningful analysis and research.

+ **Data Minimization:** Collecting and retaining only the necessary personal health data reduces the potential risks associated with holding large amounts of data.

+ **Training and Ethical Guidelines:** Educating healthcare professionals and staff about privacy and security practices, as well as providing clear ethical guidelines, can help prevent insider threats and misuse of personal health data.

+ **Transparency and Consent:** Ensuring transparency about how personal health data is collected, used, and shared, and obtaining informed consent from individuals, promotes trust and respect for privacy.

Emerging Trends in Privacy and Security

As technology continues to advance, new trends are shaping the landscape of personal health data privacy and security:

- **Blockchain Technology:** Blockchain offers a decentralized and secure system for recording and sharing personal health data, making it tamper-resistant and enhancing privacy.

- **Privacy-Preserving Techniques:** Privacy-enhancing technologies like differential privacy and homomorphic encryption allow for data analysis while preserving individuals' privacy.

- **Ethical Considerations in AI:** Ensuring ethical use of AI in healthcare, including transparency, explainability, and fairness, is crucial for protecting personal health data.

Case Study: Privacy and Security in Personal Health Monitoring Devices

Personal health monitoring devices, such as fitness trackers and smartwatches, collect vast amounts of personal health data. While these devices offer valuable insights, they also raise privacy concerns. For example, unauthorized access to this data may reveal intimate details about an individual's lifestyle and health status. To address these concerns, manufacturers should prioritize privacy in device design, implement strong security measures, and obtain informed consent from users regarding data collection and usage.

Summary

Protecting the privacy and security of personal health data is of paramount importance in a digital healthcare landscape. Maintaining privacy and ensuring appropriate security measures enhance patient trust, autonomy, and the overall ethical conduct of healthcare providers and organizations. Adhering to ethical principles, complying with legal frameworks, and employing robust strategies can help mitigate the challenges and risks associated with the use of personal health data. As technology advances, continued vigilance, adaptability, and ethical considerations are necessary to safeguard personal health data in an ever-changing environment.

Ethical Considerations in Direct-to-Consumer Genetic Testing

Direct-to-consumer genetic testing is a rapidly growing industry that allows individuals to access information about their genetic makeup and potential health risks directly, without the involvement of healthcare professionals. These tests provide insights into ancestry, genetic traits, and susceptibility to certain diseases. While this accessibility and convenience have their benefits, it also raises important ethical considerations that need to be addressed.

Informed Consent and Autonomy

One of the primary ethical considerations in direct-to-consumer genetic testing is the issue of informed consent. Informed consent is the process through which individuals are fully informed about the nature, purpose, and potential consequences of a medical test or procedure before agreeing to it. In the context of genetic testing, informed consent becomes crucial because it involves accessing deeply personal and potentially sensitive information.

Companies offering direct-to-consumer genetic testing should ensure that individuals are provided with accurate and comprehensive information about the limitations and risks associated with the test. This includes explaining the potential psychological impact of the results, as well as the limitations of the test in predicting complex diseases. It is essential to emphasize that genetic testing alone cannot provide a definitive diagnosis or guarantee future health outcomes.

Furthermore, individuals should have the autonomy to decide whether they want to undergo genetic testing and have control over their genetic information. Companies should make it clear that participation in genetic testing is entirely voluntary, and individuals should have the choice to opt-out or withdraw their consent at any time. Safeguards must be put in place to protect the privacy and confidentiality of the genetic information obtained.

Accuracy and Interpretation of Results

The accuracy and reliability of direct-to-consumer genetic testing are critical ethical considerations. The effectiveness of these tests depends on various factors, including the quality of the testing methods used, the size and diversity of the genetic database, and the interpretation of the results. Companies should provide transparent information about the scientific evidence supporting the tests they offer, the methodology used, and the limitations of the tests.

Genetic information can be complex and challenging to interpret accurately. Individuals may receive results indicating an increased risk for certain diseases or conditions, creating anxiety and fear. It is essential for companies to provide clear and understandable explanations of the results, including the level of certainty associated with the risk estimates. Genetic counselors or healthcare professionals should be readily accessible to provide expertise and support in interpreting the results and making informed decisions based on them.

Psychological and Emotional Impact

Direct-to-consumer genetic testing can have a significant psychological and emotional impact on individuals. The results of these tests may reveal unexpected information, such as a predisposition to a serious illness or the discovery of non-paternity. This information can lead to feelings of anxiety, depression, guilt, or even discrimination.

Companies must provide adequate pre- and post-test counseling to help individuals understand and cope with the potential psychological implications of the results. This could include offering access to genetic counselors, mental health professionals, or support groups. Ensuring individuals have the necessary resources to process the information and make informed decisions is vital to mitigate any negative psychological effects.

Genetic Privacy and Data Security

The privacy and security of genetic information are central ethical considerations in direct-to-consumer genetic testing. Genetic data is highly personal and unique to each individual, containing valuable information about their health, ancestry, and even their relatives. Companies must have robust security measures in place to protect this sensitive information from unauthorized access or misuse.

It is important for individuals to understand how their genetic data will be stored, who will have access to it, and how it will be used. Companies should have clear and transparent data privacy policies, ensuring that individuals have control over their genetic information and giving them the option to decide how their data is shared or used for research purposes. Consent should be sought explicitly for any secondary use of the data, such as sharing it with third parties or conducting further research.

Regulation and Oversight

Given the rapid growth of the direct-to-consumer genetic testing industry, it is essential to establish appropriate regulatory frameworks to ensure ethical practices. Regulatory bodies should impose standards for accuracy, transparency, informed consent, privacy, and security. Companies offering genetic testing services should be accountable for adhering to these regulations, with appropriate penalties for non-compliance.

Collaboration between regulatory bodies, healthcare professionals, and genetic counselors is crucial to develop guidelines that protect individuals' interests and promote responsible and ethical use of genetic testing. Ongoing monitoring and evaluation of the industry are necessary to identify emerging ethical challenges and adapt regulations accordingly.

Example: The Impact of Direct-to-Consumer Genetic Testing

To illustrate the ethical considerations in direct-to-consumer genetic testing, let's consider an example. Sarah, a 35-year-old woman, decides to take a direct-to-consumer genetic test out of curiosity about her ancestry and potential health risks. The test reveals that she has a genetic variant associated with an increased risk of breast cancer.

Upon receiving this information, Sarah experiences heightened anxiety and becomes preoccupied with thoughts of developing breast cancer. She starts to research extensively about preventive measures, even considering prophylactic surgery. However, she lacks the necessary understanding of the actual risk associated with the variant and the importance of considering other factors, such as family history or lifestyle factors.

In this scenario, the ethical considerations discussed earlier come into play. Sarah's informed consent should have included a clear understanding of the limitations and complexity of the test results. Genetic counseling should have been readily available to help Sarah interpret the results accurately and make informed decisions about her health. Proper guidance and support could have prevented unnecessary anxiety and helped her navigate the complex landscape of preventive measures.

Conclusion

Direct-to-consumer genetic testing offers exciting opportunities for individuals to access information about their genetic makeup and potential health risks. However, it also raises important ethical considerations regarding informed consent, accuracy and interpretation of results, psychological impact, genetic privacy and data security, and regulation and oversight. By addressing these considerations, we can ensure the responsible and ethical use of direct-to-consumer genetic testing, empowering individuals to make informed decisions about their health and wellness.

Implications of Personal Health Data Ownership

As the global health and wellness landscape continues to evolve, the collection and use of personal health data have become increasingly prevalent. With the advent of advanced technologies and the widespread use of digital platforms, individuals are generating large amounts of health-related data on a daily basis. This data includes information about their medical history, lifestyle choices, genetic makeup, and even real-time physiological measurements. While the availability of such data holds great promise for improving healthcare outcomes and enabling personalized treatments, it also raises important ethical considerations, particularly regarding ownership and control of personal health data.

In this section, we will explore the implications of personal health data ownership, discussing the ethical, legal, and social aspects that arise in this context. We will consider the challenges related to data privacy, security, access, and the potential benefits and risks associated with the ownership and use of personal health data.

Data Ownership and Control

When it comes to personal health data, questions of ownership and control become crucial. On one hand, individuals generate this data through their daily activities, and it reflects their personal experiences and health conditions. As such, individuals may argue that they have a right to own and control their health data. This viewpoint aligns with the autonomy principle in healthcare ethics, which emphasizes respect for individuals' rights and their ability to make choices regarding their health.

On the other hand, healthcare providers, researchers, and technology companies argue that they also have a stake in personal health data. Providers may argue that they need access to comprehensive health data to provide accurate diagnoses and appropriate treatments. Researchers may argue that access to large datasets can facilitate scientific advancements and lead to improved health outcomes. Technology companies may argue that they can develop innovative digital tools and solutions based on personal health data, leading to improved wellness experiences for individuals.

Data Privacy and Security

One of the primary concerns associated with personal health data ownership is privacy and security. The sensitive nature of health information necessitates robust safeguards to protect individuals' privacy rights and mitigate the risk of data breaches and unauthorized access.

In many countries, specific regulations and laws govern the storage, transmission, and use of personal health data. For example, the Health Insurance Portability and Accountability Act (HIPAA) in the United States sets standards for protecting individuals' health information and mandates that healthcare organizations implement safeguards to ensure privacy and security.

However, the increasing digitization and interconnectedness of health systems and technologies pose unique challenges to maintaining data privacy and security. Cyberattacks targeting healthcare organizations have become more frequent, and the potential for data breaches is a growing concern. Additionally, the use of third-party platforms and digital health applications further complicates data privacy as these entities may have their own privacy policies and data handling practices.

Access and Use of Personal Health Data

The issue of access to personal health data is closely tied to ownership and control. Individuals often face difficulties in accessing their own health records or sharing them with healthcare providers. Fragmented healthcare systems, lack of interoperability, and restrictive health data governance policies can hinder individuals' ability to access and utilize their own health information.

At the same time, it is essential to strike a balance between individuals' control over their health data and the need for healthcare providers, researchers, and technology companies to access and use this data for the collective benefit of society. One potential solution is the development of secure and user-centric platforms that empower individuals to control their health data, decide who can access it, and even participate in research studies by providing consent for data use.

Benefits and Risks

The ownership and use of personal health data come with both potential benefits and risks. On the one hand, the availability of comprehensive and high-quality health data can lead to more accurate diagnoses, personalized treatment plans, and improved health outcomes. Researchers can also leverage large datasets to identify patterns and trends, leading to the development of new therapies and interventions.

However, the misuse or unauthorized access to personal health data can result in significant harm to individuals. Data breaches can expose sensitive health information, leading to identity theft, discrimination, or social stigmatization. Moreover, the use of personal health data for commercial purposes without individuals' knowledge or consent raises ethical concerns about exploitation and breach of trust.

Ethical Frameworks and Solutions

Addressing the implications of personal health data ownership requires the application of ethical frameworks that balance individual rights, public health interests, and technological advancements. Several principles and approaches can guide the discussion:

- **Autonomy:** Respecting individuals' autonomy rights and empowering them to control their health data.

- **Beneficence:** Ensuring that the use of personal health data promotes individual and societal well-being.

- **Privacy and Confidentiality:** Implementing robust safeguards to protect individuals' privacy and maintain data confidentiality.

- **Transparency and Informed Consent:** Ensuring individuals have access to clear information about how their health data will be used and obtaining their informed consent for data sharing.

- **Data Sharing Agreements:** Establishing clear agreements between individuals, healthcare providers, and researchers on how personal health data will be shared, with appropriate safeguards and limitations.

- **Data Governance and Regulation:** Implementing comprehensive regulations and policies that govern the collection, use, storage, and sharing of personal health data.

To address these concerns, policymakers, healthcare organizations, researchers, and technology companies must work collaboratively to develop ethical and legal frameworks that safeguard individuals' privacy while enabling the responsible and beneficial use of personal health data. Additionally, initiatives such as public awareness campaigns and education programs can empower individuals to make informed decisions about their health data and advocate for their rights.

Real-World Example: The General Data Protection Regulation (GDPR)

An example of legislation that aims to protect personal data, including health data, is the General Data Protection Regulation (GDPR) enacted in the European Union (EU). The GDPR establishes principles and rules for the processing of personal data, prioritizing individuals' rights and data protection.

Under the GDPR, individuals have the right to access their personal health data, rectify inaccuracies, and even request its erasure in certain circumstances. Organizations handling personal health data are required to implement robust data protection measures and obtain individuals' informed consent before processing their data for specific purposes. Non-compliance with the GDPR can result in significant penalties, further emphasizing the importance of data privacy and protection.

The GDPR represents an example of how regulations can be designed to address the implications of personal health data ownership, striking a balance between individuals' rights and the requirements of healthcare providers, researchers, and other stakeholders.

Conclusion

The implications of personal health data ownership are multifaceted and require careful consideration of ethical, legal, and social factors. Balancing individual autonomy, privacy, and the collective benefits of data use is a complex task that demands collaboration and dialogue across various stakeholders. By addressing the challenges related to data privacy, security, access, and control, we can foster trust, advance research, and ultimately harness the potential of personal health data to shape a more equitable and just health and wellness future.

Ethical Issues in Augmented Reality and Virtual Reality in Healthcare

Informed Consent and Safety in VR and AR Healthcare Applications

In the rapidly advancing field of virtual reality (VR) and augmented reality (AR) healthcare applications, it is crucial to address the ethical concerns surrounding informed consent and safety. As these technologies become more integrated into medical practice, it is essential to ensure that individuals are fully informed about the potential risks and benefits associated with their use.

Importance of Informed Consent

Informed consent is the process by which individuals are provided with relevant information regarding a medical intervention, enabling them to make autonomous decisions about their healthcare. In the context of VR and AR healthcare applications, informed consent plays a critical role in respecting individuals' autonomy and rights.

When seeking informed consent for the use of VR and AR in healthcare, healthcare professionals should provide comprehensive information about the purpose of the intervention, the potential risks and benefits, the expected outcomes, and any alternative treatment options. Additionally, individuals should be made aware of the limitations and potential sensory discomfort that may arise during the use of VR and AR technologies.

Informed consent in VR and AR healthcare applications should also address issues such as data privacy and security. Individuals should be informed about how their personal health data will be collected, stored, and used during the VR or AR experience, and they should have the option to provide or withhold consent for data sharing.

Ensuring Safety in VR and AR Healthcare

Safety is of paramount importance when integrating VR and AR technologies into healthcare settings. The following safety considerations must be addressed:

Physical Safety: VR and AR technologies involve the use of headsets, controllers, and other equipment that may pose physical risks, such as tripping or colliding with objects in the physical environment. Healthcare professionals must educate patients about potential hazards and provide guidance on how to ensure a safe physical environment during VR or AR experiences.

Psychological Safety: VR and AR applications have the potential to induce various psychological and emotional responses, including anxiety, dizziness, or visual discomfort. Informed consent should explicitly discuss these potential risks, and healthcare professionals should monitor patients closely for any adverse psychological effects during the experience.

Cybersecurity: In healthcare applications that involve the use of VR or AR, ensuring data privacy and security is paramount. Personal health information transmitted or stored within these technologies must be protected against unauthorized access or breaches. Healthcare providers should have robust cybersecurity measures in place to safeguard patient data.

Case Study: VR for Pain Management

To illustrate the importance of informed consent and safety in VR healthcare applications, consider the use of VR for pain management. VR has been shown to effectively reduce pain intensity and improve overall patient experience during various medical procedures, such as wound care and physical therapy.

In this scenario, prior to using VR for pain management, healthcare professionals should obtain informed consent from the patients. The consent process should include a discussion of the purpose of VR, potential benefits, possible side effects (e.g., nausea, dizziness), and precautions to ensure physical and psychological safety. Patients should have the opportunity to ask questions, express concerns, and make an informed decision about their participation.

During the VR experience, healthcare professionals should closely monitor patients for any signs of discomfort or adverse reactions. Clear instructions on how to safely use and remove VR equipment should be provided, along with guidance on maintaining a safe physical environment to avoid accidental injury.

Ethical Considerations and Challenges

Despite efforts to ensure informed consent and safety in VR and AR healthcare applications, challenges may arise. Some individuals may have difficulty comprehending the information provided or may be unable to provide valid consent due to cognitive or language barriers. In such cases, healthcare professionals should involve family members or advocates to facilitate the informed consent process and ensure the individual's best interests are upheld.

Additionally, healthcare professionals must be mindful of the potential power imbalance in the patient-provider relationship. Individuals may feel pressured to consent to VR or AR interventions to please their healthcare provider or due to a fear of receiving suboptimal care if they refuse. It is crucial to create a supportive environment that encourages individuals to freely express their concerns, ask questions, and make autonomous decisions regarding their healthcare.

Educational Resources and Guidelines

Given the evolving nature of VR and AR healthcare applications, healthcare professionals should stay informed about the latest developments, ethical guidelines, and best practices. The American Medical Association (AMA) has issued guidelines on the ethical use of telemedicine, including VR and AR technologies, which can serve as a valuable resource for healthcare providers.

Additionally, professional organizations and academic institutions may offer training programs or courses that focus on the ethical considerations and challenges posed by VR and AR in healthcare. These educational resources can equip healthcare professionals with the knowledge and skills needed to navigate the complex ethical landscape of these technologies.

Conclusion

Informed consent and safety are paramount in the use of VR and AR technologies in healthcare. By ensuring individuals have a comprehensive understanding of the potential risks and benefits and taking measures to address physical and psychological safety concerns, healthcare professionals can uphold ethical principles and promote patient autonomy. Continued research, education, and adherence to guidelines will be crucial in shaping an ethical and safe future for VR and AR healthcare applications.

Ethical Dilemmas in Using VR and AR for Therapeutic Purposes

Virtual Reality (VR) and Augmented Reality (AR) technologies have become increasingly popular in various fields, including healthcare. VR refers to an immersive experience where users are fully immersed in a computer-generated environment, while AR overlays digital information onto the real world. These technologies offer exciting possibilities in therapeutic settings by creating simulated environments that can be tailored to specific therapeutic goals. However, the use of VR and AR for therapeutic purposes also raises important ethical dilemmas that need to be addressed.

Informed Consent and Autonomy

One of the key ethical concerns in using VR and AR for therapeutic purposes is obtaining informed consent from patients. Informed consent ensures that patients have the necessary information and understanding to make autonomous decisions about their participation in therapy. However, the immersive nature of VR and AR experiences may make it challenging to fully comprehend and anticipate the range of effects and potential risks involved.

Therapists and healthcare professionals must provide clear and comprehensive information about the objectives, benefits, limitations, and potential risks associated with VR and AR therapy. This includes explaining the potential psychological and emotional responses that patients may experience during or after the therapy. It is crucial to ensure that patients fully understand the nature of the virtual or augmented experience, the purpose of the therapy, and any potential long-term effects.

Moreover, therapists must consider the capacity of patients to provide informed consent. Some patients, such as individuals with severe mental illnesses or cognitive impairments, may have limited decision-making capacity. In such cases, therapists need to involve family members or legal guardians in the decision-making process, ensuring that the patient's best interests are upheld.

Privacy and Confidentiality

VR and AR technologies collect and process sensitive personal data to create personalized therapeutic experiences. This raises concerns about privacy and confidentiality. Therapists and healthcare organizations need to ensure that patient data is securely stored, transmitted, and used only for the intended purposes.

Strict privacy protection measures should be in place to safeguard patient data. This includes using encryption techniques, access controls, and anonymization methods to minimize the risk of data breaches or unauthorized use. Therapists must inform patients about data collection practices, the purpose of data gathering, and the measures in place to protect their privacy. Patients should also have the right to control the use and sharing of their personal data.

Ethical guidelines should be established to govern the use of patient data for research and development purposes. Transparency and accountability in data usage are essential to build trust between patients, therapists, and healthcare organizations.

Therapist-Patient Relationship

The use of VR and AR technologies can alter the dynamics of the therapist-patient relationship. In traditional therapeutic settings, therapists rely on face-to-face interactions, body language, and non-verbal cues to establish rapport and build trust. However, VR and AR experiences may create a sense of detachment or artificiality that can impact the therapeutic relationship.

Therapists must be cautious not to rely solely on VR or AR technologies as substitutes for human connection. They should maintain regular check-ins with patients, providing opportunities for open dialogue and emotional support. It is crucial for therapists to remain attentive to patients' emotional states and address any concerns or conflicts that may arise during or after the VR or AR therapy sessions.

Equity and Accessibility

Another ethical consideration in using VR and AR for therapeutic purposes is accessibility and equity. These technologies may introduce additional barriers to therapy for certain populations, such as individuals with disabilities, low-income individuals, or those living in remote areas with limited access to technology.

Healthcare organizations and therapists need to ensure that the use of VR and AR technologies does not exacerbate existing inequities in access to mental health services. Efforts should be made to provide equal opportunities for all patients to benefit from these technologies, regardless of their socioeconomic status or physical abilities. This may involve providing financial support, equipment loan programs, or ensuring that therapy alternatives are available for those who may face barriers to accessing VR or AR therapy.

Professional Competence and Training

Therapists and healthcare professionals using VR and AR technologies for therapeutic purposes should possess the necessary skills and competence to ensure patient safety and effective therapy outcomes. This includes understanding the ethical implications of using these technologies and having the technical knowledge to operate them.

Healthcare organizations should provide adequate training and support to therapists, ensuring that they are competent in using VR and AR technologies and integrating them into their existing therapeutic practices. Ongoing professional development and supervision are essential to stay updated with emerging ethical issues, best practices, and safety protocols related to VR and AR therapy.

Example: Using VR to Treat Post-Traumatic Stress Disorder (PTSD)

Consider the example of using VR for the treatment of Post-Traumatic Stress Disorder (PTSD). VR can recreate traumatic scenarios in a controlled environment, allowing patients to confront their fears and anxieties in a safe setting. While this therapy has shown promising results, it also presents ethical dilemmas.

One ethical dilemma is the potential retraumatization of patients during VR exposure therapy. Patients may experience intense distress or emotional reactions when exposed to simulated traumatic events. Therapists must carefully assess and monitor patients' readiness and provide sufficient support throughout the therapy process.

Additionally, therapists need to consider the potential long-term psychological impact of VR therapy. Extended exposure to emotionally challenging scenarios may have unintended consequences, such as desensitization or exacerbation of symptoms. Proper debriefing and follow-up care are crucial in mitigating these risks.

Overall, the ethical dilemmas in using VR and AR for therapeutic purposes necessitate a thoughtful and nuanced approach. By addressing the issues of informed consent, privacy and confidentiality, therapist-patient relationship, equity and accessibility, professional competence and training, we can harness the potential of these technologies while upholding ethical standards in healthcare delivery.

Virtual Reality and the Impossibility of Experience

Virtual reality (VR) has become increasingly popular in recent years, offering an immersive and interactive experience that simulates a three-dimensional environment. It has found applications in various fields, including entertainment, education, and healthcare. However, as virtual reality technology continues to advance, it raises ethical questions about the nature of experience and its implications for human perception and reality.

The Illusion of Reality in Virtual Environments

Virtual reality creates a sense of presence, where users feel they are physically present in a computer-generated environment. Through the use of headsets, haptic devices, and motion tracking, VR can provide a highly realistic and engaging experience. However, it is important to recognize that VR is ultimately an illusion and not a true substitute for reality.

The immersive nature of virtual reality can make it difficult for users to distinguish between the virtual world and the physical world. This blurring of boundaries raises concerns about the potential negative effects on human perception and cognition. For example, prolonged exposure to VR environments may lead to a disconnection from real-world experiences and a distorted sense of reality.

Ethical Considerations in Virtual Reality

The use of virtual reality technology raises several ethical considerations. One key issue is the potential for harm to users. While VR experiences can be enjoyable and educational, they can also be intense and overwhelming. Developers and providers of VR content must prioritize the well-being of users by ensuring that experiences are safe, avoiding excessive sensory stimulation, and providing clear guidelines and warnings.

Another ethical concern is the potential for addiction to virtual reality. Like other forms of digital media, excessive use of VR could lead to dependency and withdrawal symptoms. It is important for individuals and society to recognize the importance of maintaining a balanced and healthy relationship with technology, including virtual reality.

Privacy and data security are also significant ethical considerations in virtual reality. VR systems often collect data about users, including their movements, behaviors, and preferences. This data can be valuable for improving user experiences but must be handled with care to protect individual privacy and prevent misuse.

The Philosophical Puzzle of Experience

Virtual reality poses a challenge to our understanding of subjective experience. The question of whether virtual experiences are qualitatively the same as real experiences raises fundamental philosophical questions about consciousness, perception, and the nature of reality itself.

One philosophical puzzle is the problem of other minds. In virtual reality, we may interact with virtual beings that have apparent thoughts, emotions, and consciousness. However, it is unclear whether these virtual entities truly have subjective experiences or are just simulating them. The philosophical debate surrounding the nature of consciousness and the possibility of artificial consciousness has important implications for the ethics of creating and interacting with virtual beings.

Additionally, virtual reality challenges our notions of authenticity and genuine experience. While virtual experiences can be immersive and enjoyable, some argue that they are fundamentally different from real experiences. The feeling of presence in virtual reality may not be equivalent to true presence in the physical world. This distinction raises questions about the value and significance of virtual experiences and their impact on our understanding of what it means to be human.

Balancing Virtual and Real-world Experiences

While virtual reality offers unique and exciting possibilities, it is important to strike a balance between virtual experiences and real-world interactions. Virtual reality should not replace or overshadow the richness of physical experiences, relationships, and the natural world.

As individuals and as a society, we must be mindful of the potential consequences of excessive reliance on virtual reality. Maintaining a healthy balance between virtual and real-world experiences involves setting boundaries, prioritizing real-life interactions, and taking breaks from immersive technologies.

Summary

Virtual reality presents a range of ethical issues, including the illusion of reality, potential harm to users, addiction, privacy concerns, and philosophical puzzles about the nature of experience. While virtual reality can offer valuable and engaging experiences, it is crucial to approach it with caution and balance. Striking a balance between virtual and real-world experiences is essential for maintaining our connection with reality and understanding the true depth of human experience.

Conclusion: Towards an Ethical and Just Health and Wellness Future

The Role of Ethics in Shaping Health and Wellness Policy

Ethical Considerations in Policy Development and Implementation

Policy development and implementation in the context of health and wellness play a crucial role in shaping the ethical landscape of the field. Effective policies can have a significant impact on the equitable distribution of resources, the promotion of justice, and the overall well-being of individuals and communities. However, policy decisions often involve complex ethical dilemmas and trade-offs. In this section, we will explore some key ethical considerations that need to be taken into account when developing and implementing health and wellness policies.

Balancing Individual Autonomy and the Common Good

One of the fundamental ethical considerations in policy development is the balance between individual autonomy and the common good. Health and wellness policies are intended to promote the well-being of individuals and communities as a whole. However, these policies may sometimes infringe upon individual autonomy and personal choices. For example, policies related to mandatory vaccination or restrictions on certain behaviors may limit individual freedom in the interest of public health.

To address this ethical dilemma, policymakers must carefully consider the extent to which individual autonomy should be restricted for the greater good. Transparency and public engagement are crucial in this process. It is important to involve stakeholders, such as healthcare providers, advocacy groups, and community members, in policy discussions to ensure that a wide range of perspectives are considered. Furthermore, clear justifications and evidence-based reasoning should be provided to explain why certain restrictions on individual autonomy are necessary.

Ensuring Equity and Justice

Another critical ethical consideration in policy development and implementation is the promotion of equity and justice. Health and wellness policies should aim to address health disparities and inequalities, particularly among marginalized and vulnerable populations. Policies that perpetuate or exacerbate existing disparities are ethically problematic and need to be carefully examined.

To ensure equity, policies need to take into account the social determinants of health and wellness. For example, policies that prioritize access to healthcare services in underserved areas or provide financial assistance for low-income individuals can help address socioeconomic disparities. Additionally, policies need to consider the unique needs and perspectives of different populations, including racial and ethnic minorities, LGBTQ+ individuals, and people with disabilities.

Furthermore, justice should be at the forefront of policy considerations. Policies should be based on principles of fairness and impartiality, taking into account the interests and well-being of all individuals and communities. This includes considering the needs of future generations by implementing sustainable practices and policies that protect the environment and promote long-term health and wellness.

Ethical Decision-making and Accountability

Ethical decision-making processes and mechanisms for accountability are essential in policy development and implementation. Policies should be based on evidence, research, and ethical principles. Informed by these factors, policymakers should consider multiple perspectives to ensure that potential conflicts or unintended consequences are addressed.

Transparency is key in ensuring accountability. Policymakers should communicate the rationale behind policy decisions and provide opportunities for public input. Additionally, the decision-making process should be transparent, with clear documentation of the evidence considered and the reasoning behind the chosen policy.

Evaluation and monitoring of policies are crucial to ensure their effectiveness and ethical soundness. Regular assessments should be conducted to determine whether policies are achieving their intended goals and whether any unintended negative consequences have emerged. If a policy is found to be ineffective or unethical, adjustments should be made to rectify the situation.

Addressing Power Imbalances

Power imbalances can significantly impact policy development and implementation, leading to inequities and injustices. Policymakers need to be aware of these power dynamics and take steps to address them.

Engaging diverse stakeholders and consulting with communities affected by the policies can help mitigate power imbalances. Involving those with lived experiences, community organizations, and advocacy groups can provide valuable insights and help ensure that policies address the needs of all individuals.

Similarly, promoting transparency and accountability can help counteract the influence of powerful interest groups or individuals. By making the decision-making process accessible and accountable, policymakers can ensure that policies are developed and implemented in a fair and equitable manner.

Educating and Empowering Individuals

Policy development and implementation should not solely rely on top-down approaches. Educating and empowering individuals is an essential ethical consideration in promoting health and wellness. Policies should be accompanied by educational initiatives that inform individuals of their rights and responsibilities, as well as the rationale behind specific policies.

Empowering individuals to make informed choices and take an active role in their health and wellness can contribute to more ethical and effective policy outcomes. This can be achieved through health literacy programs, community engagement initiatives, and the provision of accessible and understandable information.

By ensuring that individuals are equipped with the knowledge and resources to make informed decisions, policymakers can promote autonomy, equity, and justice in health and wellness.

Conclusion

Ethical considerations play a crucial role in the development and implementation of health and wellness policies. Balancing individual autonomy and the common good, ensuring equity and justice, promoting ethical decision-making and accountability, addressing power imbalances, and educating and empowering individuals are all essential aspects of ethical policy development.

By carefully considering these ethical considerations, policymakers can contribute to the creation of policies that are not only effective but also promote justice, equity, and a sustainable and equitable health and wellness future. It is imperative to continually reflect on and evaluate the ethical dimensions of health and wellness policies to ensure that they align with the core values of the field.

Strategies for Promoting Ethical and Just Health and Wellness Systems

Promoting ethical and just health and wellness systems is crucial for ensuring that individuals have access to high-quality and equitable healthcare services. In this section, we will explore several strategies that can be implemented to achieve this goal. These strategies highlight the importance of ethical decision-making, social justice, and sustainability in the delivery of health and wellness services.

Ethical Decision-Making

Ethical decision-making forms the foundation of promoting ethical and just health and wellness systems. Healthcare providers should be equipped with the necessary knowledge and skills to engage in ethical deliberation and resolve moral dilemmas that arise in clinical practice. To promote ethical decision-making, the following strategies can be implemented:

1. **Ethics Education and Training:** Healthcare professionals should receive comprehensive ethics education and training. This training should include discussions on ethical theories, principles, and frameworks, as well as case studies and role-playing exercises to enhance their ability to make ethically informed decisions.

2. **Ethical Guidelines and Protocols:** Healthcare organizations should develop and implement clear ethical guidelines and protocols that outline the principles and values that govern the delivery of care. These guidelines should address various

ethical issues, such as end-of-life care, resource allocation, and informed consent, providing healthcare providers with a framework to guide their decision-making.

3.   **Ethical Consultation and Support:** Healthcare organizations should establish mechanisms for ethical consultation and support, such as ethics committees or clinical ethics consult services. These resources can provide guidance to healthcare professionals facing ethically complex situations and help them navigate moral dilemmas.

Social Justice

Promoting social justice is essential for creating a more equitable and just health and wellness system. This involves addressing systemic barriers and inequalities that prevent certain populations from accessing quality healthcare. The following strategies can be employed to promote social justice:

1. **Health Equity Assessments:**   Healthcare organizations should conduct regular health equity assessments to identify disparities in healthcare access and outcomes. These assessments can help pinpoint areas of improvement and guide the allocation of resources to address the health needs of underserved populations.

2. **Culturally Competent Care:**   Healthcare providers should strive to deliver culturally competent care that recognizes and respects the diverse backgrounds, values, and beliefs of their patients. This involves developing an understanding of cultural norms, language barriers, and healthcare disparities that may affect different populations, and tailoring care accordingly.

3. **Community Engagement:**   Healthcare organizations should actively engage with the communities they serve to ensure that health and wellness initiatives are responsive to their needs. This can involve partnerships with community-based organizations, conducting outreach programs, and involving community members in the design and implementation of healthcare interventions.

Sustainability

Sustainability in health and wellness systems involves ensuring the long-term availability and accessibility of healthcare resources, while minimizing the impact on the environment.   The following strategies can be adopted to promote sustainability:

1. **Environmental Stewardship:** Healthcare organizations should prioritize environmentally friendly practices by implementing energy-efficient technologies, reducing waste generation, and promoting recycling and sustainable procurement. This can contribute to reducing the carbon footprint of healthcare facilities and minimizing the negative impact on the environment.

2. **Preventive Care and Health Promotion:** Emphasizing preventive care and health promotion can significantly reduce the burden on healthcare systems. By focusing on education, early detection, and lifestyle interventions, individuals can be empowered to take control of their health, leading to better health outcomes and reduced healthcare costs in the long run.

3. **Collaboration and Resource Sharing:** Collaboration among healthcare organizations can promote resource sharing and optimize the use of healthcare resources. This can involve sharing best practices, pooling resources, and coordinating efforts to ensure equitable access to healthcare services across different regions.

In conclusion, promoting ethical and just health and wellness systems requires a multi-faceted approach. Ethical decision-making, social justice, and sustainability must be prioritized to ensure that healthcare services are accessible, equitable, and of high quality. By implementing the strategies outlined in this section, we can work towards a future that promotes the well-being of individuals and communities while striving for a more ethical and just healthcare system.

Achieving a Sustainable and Equitable Health and Wellness Future

In order to achieve a sustainable and equitable health and wellness future, it is crucial to address the underlying factors that contribute to health disparities and inequities. This section will explore key strategies and approaches to promote a more just and equitable healthcare system, as well as ways to ensure the sustainability of health and wellness initiatives.

Addressing Social Determinants of Health

One of the fundamental steps in achieving a sustainable and equitable health and wellness future is to address the social determinants of health. These are the social, economic, and environmental factors that influence health outcomes and contribute to health disparities. By focusing on these determinants, we can effectively target the root causes of health inequities.

To address social determinants, it is essential to implement policies and interventions that tackle issues such as poverty, unemployment, inadequate housing, and food insecurity. For example, initiatives aimed at reducing income inequality and providing affordable housing can have a significant impact on improving health outcomes. Additionally, promoting access to healthy food options and ensuring safe and clean environments can contribute to better overall health and wellness.

Promoting Health Literacy and Education

Another key aspect of achieving a sustainable and equitable health and wellness future is promoting health literacy and education. Health literacy refers to the ability of individuals to access, understand, and apply health information in order to make informed decisions about their own health. By improving health literacy, we can empower individuals to take control of their health and make choices that promote well-being.

Efforts to promote health literacy can include providing accessible and culturally sensitive health information, enhancing communication between healthcare providers and patients, and promoting health education in schools and communities. By equipping individuals with the knowledge and skills to make informed decisions about their health, we can reduce health disparities and promote equitable access to healthcare.

Advancing Technology and Innovation

Technology and innovation have the potential to revolutionize the delivery of healthcare services and promote sustainability. Advances in telemedicine, artificial intelligence, and digital health solutions can help overcome barriers related to distance, cost, and accessibility.

Telemedicine, for example, allows individuals to access healthcare services remotely, reducing the need for in-person visits and increasing access to care, particularly in underserved areas. Additionally, artificial intelligence can enhance diagnostic accuracy and treatment effectiveness, leading to improved health outcomes. By embracing technological advancements and fostering innovation, we can bridge the gap in healthcare access and promote equity.

Engaging Communities in Decision-making

To achieve a sustainable and equitable health and wellness future, it is crucial to involve communities in decision-making processes. Community engagement ensures that the diverse needs and perspectives of the population are taken into consideration. By actively involving community members in the planning, implementation, and evaluation of healthcare policies and programs, we can promote equity and address the specific needs of different populations.

Community engagement can take various forms, including community-based participatory research, community health forums, and advocacy groups. By fostering collaboration and partnerships between healthcare providers, policymakers, and community members, we can create a healthcare system that is responsive, inclusive, and equitable.

Ensuring Sustainable Healthcare Financing

Sustainable healthcare financing is essential for achieving a sustainable and equitable health and wellness future. It is important to ensure that the necessary resources are allocated efficiently and effectively to meet the healthcare needs of the population.

One approach to sustainable healthcare financing is the implementation of universal healthcare systems that provide equitable access to healthcare services for all individuals. By pooling resources and redistributing them based on need, universal healthcare systems can help address health disparities and promote equity. However, it is also important to consider the sustainability of healthcare financing models and explore innovative financing mechanisms to ensure the long-term viability of healthcare systems.

Conclusion

Achieving a sustainable and equitable health and wellness future requires a comprehensive and multi-faceted approach. By addressing social determinants of health, promoting health literacy and education, embracing technology and innovation, engaging communities in decision-making, and ensuring sustainable healthcare financing, we can work towards a future where everyone has equal opportunities to lead healthy and fulfilling lives. It is through these collective efforts that we can shape a more just and equitable world, where health and wellness are accessible to all.

Index

Milton Keynes UK
Ingram Content Group UK Ltd.
UKHW020845051124
450766UK00013B/914